So_p

Early History of the Nuclear Atom

Early History
of the
Nuclear Atom

William E. Davis, Jr.
Boston University

KLH

KENDALL/HUNT PUBLISHING COMPANY
DUBUQUE, IOWA

Copyright © 1974 by William E. Davis, Jr.

Library of Congress Catalog Card Number: 74—82179

ISBN 0—8403—0973—2

Printed in the United States of America

To My Parents

Contents

Preface

This book presents a history of the development of man's ideas of the nature of the atom from about 1850 to 1932. There is a heavy reliance on edited portions of the original scientific publication of these ideas. In this preface, I wish first to deal with the particular problems involved in preparing an edited work, and second, to discuss why I think books of this type are useful and necessary.

I was, by necessity, forced to choose, not only what papers were to be included in this volume, but also what sections of each paper should be included, and in addition, their sequence of presentation. This necessarily introduced several of my biases. I have chosen, in an attempt to insure clarity, to include only those parts of each original paper which seem to me to support the general theme. I have thus deleted, for the most part, those portions of the original papers which retrospect has indicated were either in serious error or superfluous. This tends to produce the impression that the original papers and their authors were "more correct" and demonstrated more clarity of thought than was actually the case. Even many of Rutherford's papers contained information, the value of which has been diminished by subsequent discovery and analysis.

An editor, in attempting to present any historical picture, must always select but a few meagre samples from an incredible wealth of historical material, and in doing so reflects his own biases of training, environment and philosophy. He always presents his own particular view of history.

The hazards of selectively editing a paper include the possibility of distorting the author's work, or misrepresenting the author's intentions.

Any presentation less than the entire work is to some degree a misrepresentation. Most selective editing tends to simplify the original presentations and problems, as the more complex data and irrelevant material tend to be edited out. This can lead to an oversimplification where the complex nature of the original problem is completely ignored.

By dividing the presentation into distinct topics (e.g., Isotopes; The Discovery and Isolation of the Electron) the editor is implying a categorization which historically, in fact, did not exist. All the researches were to a greater or lesser degree interrelated and interdependent, with no one man or group of men working on just isotopes or just the electron (particularly before the concept of isotopes or the electron were formulated). Without these somewhat arbitrary categorizations however, it would be very difficult to present an historical picture in other than a disorganized manner.

It is hoped that in this volume all of these editing difficulties have been kept to a minimum.

There are a number of reasons why I think that edited books of this type are both worthwhile and necessary.

There appears to be an increasing demand on the part of college students of all levels to examine primary sources rather than textbook or other secondary accounts. I think this is a very healthy trend.

Primary sources give the reader some indication of the social environment in which the original work was done, and some indication of the personality of the worker through the style and content of his published work. This is particularly important if the reader is examining the historical development of ideas. No secondary report of Rutherford's work, for example, can convey the clarity and forcefulness of his presentations as well as these works themselves.

Textbooks in Physical Science, and introductory texts in Chemistry and Physics generally give too brief and superficial accounts of any specific area of thought because they attempt to cover such vast areas of human knowledge. This is certainly true in the area of the historical development of man's ideas concerning the atom. Therefore, there is a need for supplementary works which are more specific in their focus, and which by use of extensive primary sources, present a fuller picture of the historical setting in which these ideas were developed.

It is hoped that this book fulfills this function. It is concerned with the development of man's ideas of the atom from about 1850 to 1932, primarily restricted to the context of experimental physics. This is in no way meant to imply that the stupendous advances in theoretical physics during this period by individuals such as Einstein, Plank, Heisenberg, Dirac and many others, were not important in developing our ideas concerning the atom, for they most certainly were important. Several considerations, however, dictated the more limited scope of this book. First, the inclusion of many theoretical works would have exceeded the space limitations of this book. Second, it is hoped that this book can be read and understood by people with very limited mathematical backgrounds. It would be very difficult to include the many theoretical advances which were largely dependent on a sophisticated level of mathematics, and still make the book readable for individuals with limited math skills.

The period of 1850 to 1932 in the history of experimental science is particularly interesting. It is classic in the sense that the experimental science of this period resembled the textbook picture of science. The step-by-step development of the model of the atom clearly exemplifies the cumulative nature of science. Many of the experiments performed during this period, and much of the thinking involved, demonstrate the systematic approach to the accumulation of knowledge generally referred to as "the scientific method."

The momentous impact on civilization during the past few decades which knowledge of the atom has effected, makes the accumulation of this knowledge relevant to everyone today.

The unraveling of the "mysteries" of the atom has a detective story element to it, which gives the study of this period of experimental science an unusually entertaining quality.

In conclusion, I would like to express my thanks to all those who made this book possible. In particular, I would like to thank George LeSuer, Phil Fogg, Colin Kerr, and Colin Livesey for their help and suggestions, Barbara Boughman who typed the manuscript, and Jean Allaway who was so helpful at the Library of Congress. I wish to thank also Louise Z. Smith and James D. Donovan who translated those excerpts which were originally in French.

Special thanks is due my wife Betsy who critically read the manuscript and helped throughout its preparation.

I wish to acknowledge the journals listed below for their permission to reproduce the excerpts and diagrams used in this book. The written portions of these articles have been retyped, but the diagrams are reproductions from the originals.

Comptes rendus
Nature
The Philosophical Magazine
Physical Review
Proceedings of the Royal Society

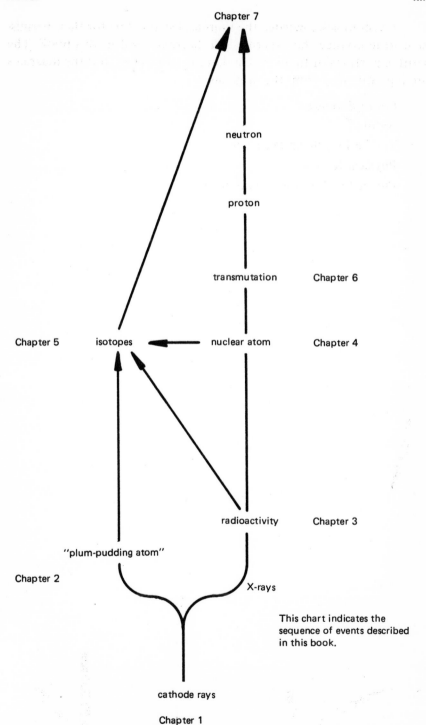

Chapter 7

neutron

proton

transmutation   Chapter 6

Chapter 5   isotopes   nuclear atom   Chapter 4

radioactivity   Chapter 3

"plum-pudding atom"

Chapter 2   X-rays

This chart indicates the
sequence of events described
in this book.

cathode rays

Chapter 1

Chapter 1

Introduction

Man's curiosity and concern about the nature of matter dates back at least to the 6th century B.C., and probably much earlier. The earliest speculations concerned an "ultimate reality," or "ultimate substance," of which all things of the world were made.

The fundamental question of whether matter is continuous or discontinuous, was under consideration at least as early as the 5th century B.C. Matter must be either continuous or discontinuous. If matter is continuous the universe must be full of matter. This denies the existence of empty space, and implies that a chunk of matter can be subdivided into smaller and smaller pieces indefinitely, without ever reaching the point where further subdivision would alter the properties of the substance. If matter is discontinuous empty space must exist, and matter can only be subdivided so far, after which no further subdivision could take place. This implies that matter is ultimately made up of particles separated by empty space. .

The preponderance of early philosophical thought, defended by such able thinkers as Aristotle, favored a continuous view of matter. The major defense of this position generally, was that since the philosophers could not conceive of empty space, empty space could not exist. Most of our sense observations tend to indicate a world full of matter, and this may have influenced some thinkers.

In the 5th century B.C., we find the first recorded speculations concerning the discontinuous nature of matter. Leucippus and his pupil Democritus speculated that the world was constructed of invisible and indivisible particles, moving about in empty space. They called these particles "atoms" (from the Greek phrase meaning indivisible), a name which is still used today. The atoms were supposed to be of different sizes and shapes and all things were constructed from combinations of these atoms. The mechanistic implications of this outlook on the nature of matter made the philosophy distasteful to most. However, we find through the ancient world an occasional supporter of this position (e.g., the Roman Lucretius).

As we trace the history of these ideas into more modern times, we find a belief in the discontinuous nature of matter with its atoms and empty space, held by such giants of science as Robert Boyle and Isaac Newton.

An examination of these early atomic speculations and beliefs shows that all of these speculations were *qualitative* in nature. There was no attempt to systematically analyze the possible atomic nature of matter from a *quantitative* point of view. The atomic theory, in other words, had not been tested either by explaining empirical observations or by making testable predictions about matter.

The real origins of our modern atomic theory date from the early years of the 19th century, when John Dalton introduced the first systematic quantitative evidence for the atomic nature of matter.

Essentially, what Dalton did was to formulate a series of assumptions about the nature of matter. Then, with these assumptions, he explained one already formulated empirical "law," and predicted another! It is worthwhile at this point to enumerate Dalton's assumptions and examine what observations they are able to explain, since these assumptions, constituting the first quantitatively based atomic theory, were almost universally accepted by scientists by the years 1860-1880 when the first works described in this book were being conducted.

Dalton's major assumptions summarized are:

(1) All matter is made up of invisible, indivisible and indestructible particles (for which the word "atom" was retained), surrounded by empty space.

(2) The atoms of an element are identical in weight and properties, and the atoms of different elements are different in weight and properties.

(3) Chemical combination takes place between the atoms of different elements to form compound atoms (molecules), and in a homogenous compound all the molecules are identical.

The empirical "law" which Dalton was able to explain is usually referred to as the Law of Definite Proportions or the Law of Fixed

Composition. It had been demonstrated in the latter part of the 18th century that in the formation of a single compound the combining elements always combine in a fixed ratio by weight. This would be difficult to explain if matter were continuous. But if matter is discontinuous, if a single compound contains identical molecules, and all the atoms of each element are identical, then the combining weights of each element in a single compound would have to be in a fixed ratio!

Dalton went further and predicted, on the basis of his assumptions, that when atoms of several elements form more than one compound, there will be a small whole number ratio in the differing weights of one of the elements in the compounds. Chemical analysis quickly demonstrated the correctness of this prediction.

The important thing to see here is that Dalton tested his atomic theory, his assumptions, in a *quantitative* manner.

One of the fascinating things about scientific discovery is that each new advance usually poses more problems than it solves. Dalton's atomic theory is no exception. One of the major problems prompted by Dalton was the determination of the *relative* weight of one atom of one element to one atom of another. There was really no way in which Dalton, with his singular approach of weighing things, could determine relative atomic weights. It was an entirely different approach to the study of matter that ultimately led to an accurate determination of relative atomic weights of the elements. This new approach involved a quantitative study of chemically combining volumes of gases.

Although a series of assumptions by a man named Avogadro very early in the 19th century had, in essence, solved the problem of determining relative atomic weights, it was not until a very persuasive gentleman named Cannizzaro convinced the scientific world of the correctness of Avogadro's approach around 1860, that accurate relative atomic weights were determined.

The determination of the relative atomic weights of the elements was of critical importance to further discoveries concerning the atom, as accurate relative atomic weights formed the basis of Mendeleev's work about 1870. He found that the chemical properties of an element seemed to depend on their atomic weight. He found that if he lined up all the elements in a row by increasing atomic weight there was a periodic reoccurrence of similar chemical properties. He modified his lineup of elements by increasing atomic weight by putting the periodically chemically similar elements below one another. He was thus able to construct the forerunner of the modern periodic table of the elements found in every chemistry textbook.

There was a very important implication drawn from the periodic table of Mendeleev. The chemical properties of the various elements were not random, but systematically varied with atomic weight. The elements must somehow be related. Even the combining power of elements (valence) seemed related to atomic weight. All this could not be explained by chance.

Hence, once again, an advance in science prompted more questions than it answered. How can one explain the periodic reoccurrence of chemical properties as a function of atomic weight? How does atomic weight control the combining power of elements? Why do some of the horizontal rows in the periodic table contain more elements than others? Are there a limited number of elements? If so, why? These are just a few of the questions raised by Mendeleev's table. One wonders how many scientists of that period suspected that the simplest way to explain the interrelated nature of the elements was through common structural features. This line of thought may have prompted the first inklings that the atom might not be indivisible, but contain some sort of subatomic structure and subatomic parts.

Once again we are at a point where further progress appeared blocked. The answers to the questions posed by the periodic table were not to be found through the means which had been employed to develop the table: the measurement of weights and volumes of elements and compounds. Instead a very different approach to the study of matter provided the first breakthroughs which led to our modern atomic theory and the answers to the questions raised by the periodic table.

This approach involved the electrical nature of matter. The history of the accumulation of our knowledge of electricity is an interesting topic in itself, but for the purpose of establishing the setting for this book, only the work of one man, Michael Faraday, need be mentioned. Faraday's work, culminating in the publication in the 1830s of his laws of electrolysis, provided the major background work necessary for the projects to be discussed in the next chapter.

Of particular importance was Faraday's establishing the fundamental interrelationship of chemical bonding and the electrical nature of matter. He found, for example, that chemical activity can produce electricity (battery) and that electricity can produce chemical activity (electrolysis). Faraday established empirically that in electrolysis a specific quantity of electricity will always release a specific weight of an element from a compound. If all matter is made up of atoms, then it follows that a specific amount of electricity is needed to release each atom. Further, it takes twice as much electricity to release an atom of a divalent element, and three times as much for a trivalent element, as it takes to release an atom of a monovalent element. This certainly suggests that electricity comes in specific amounts or packages. If so, then perhaps each package of electricity is associated with a particle of matter. Faraday never suggested this, although one can infer from his writings that he recognized the possibility. Rutherford summarized this very nicely in 1922 (46).

> The first definite proof of the close relations that exist between electricity and matter we owe to the famous experiments of Faraday on the passage of electricity through electrolytes. It was clear that the simple numerical relations found by him between the electrochemical equivalents of the elements and their atomic weights could be simply interpreted by assuming that electricity was atomic in character and that the charges carried by the individual ions were integral multiples of a fundamental unit of charge.

In this introductory chapter we have taken a brief look at the state of scientific knowledge when the first experiments were being performed which led, in the latter part of the 19th and early 20th centuries, to our modern atomic theory of the nature of matter.

Chapter 2

Discovery and Isolation of the Electron

The experiments which culminated in the discovery of the electron involved the study of the passage of electrical currents through gases at low pressure. Early studies of this sort were conducted by Faraday and others during the first half of the 19th century, but it was not until the mid-1850s, when equipment was constructed that made possible the confinement of gases at very low pressures, that experimental breakthroughs began. A glass blower named Von Geissler was largely responsible for improving the experimental equipment then available to scientists. He developed a vacuum pump which allowed him to reduce gas pressure in a tube to much lower pressures than had previously been possible. He also sealed within the tube two metal plates which could be connected to an electrical power source (the battery had been developed at the end of the 18th century).

Perhaps a few words about these tubes is in order. The plate connected to the negative terminal of the battery, following the terminology of Faraday, is called a cathode. The plate connected to the positive terminal is termed the anode (see below).

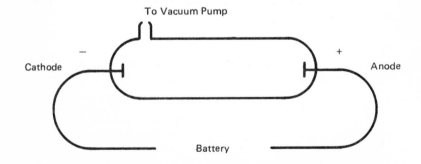

If a potential is established between the cathode and anode by their connection to a battery, and if the gas in the tube is gradually pumped out, eventually the remaining gas will begin to glow. Each different gas will glow with its own characteristic color. A dark space will develop around the cathode and a series of alternating color bands and dark spaces occupy the rest of the tube. As evacuation continues, the dark space around the cathode expands until it fills the entire tube. At this point the glass at the end of the tube near the anode begins to glow. These observations were the starting point for a series of workers, particularly in Germany, during the 1850s, 60s and 70s. They established that there were some sort of rays emitted from the cathode which traveled across the tube to the anode and named them cathode rays. Hence, the experimental apparatus is generally referred to as a cathode ray tube.

A very fine series of experiments were conducted by Sir William Crookes of England during this period, and the following excerpts from a lecture he delivered in 1879 (13) contain a discussion of these experiments. The purpose of these experiments was to determine the nature of cathode rays. The most important question he answered was: are cathode rays some sort of electromagnetic radiation (similar to light), or some sort of stream of charged particles? He built a case of overwhelming evidence in favor of the latter.

In the first few excerpts Crookes demonstrated that the rays are emitted from the cathode and travel to the anode. The cathode used in his experiment was cup-shaped, and therefore, focused the rays at a point.

GREEN PHOSPHORESCENT LIGHT OF MOLECULAR IMPACT

When the exhaustion approaches 30 M,* a new phenomenon makes its appearance. . . .

. . .At the part of the bulb on which the rays impinge, a faint spot of greenish-yellow light is observed, sharp in outline. On exhausting to 14 M, and making the cup the negative pole** of the coil, the projection from the cup is represented by a brilliant green spot of light about 7 millims. diameter, . . .

. . .On reversing the poles and making the cup positive, the bulb becomes beautifully illuminated with greenish-yellow light. . . .

The greenish-yellow phosphorescence of the soft German glass

*M signifies the millionth of an atmosphere.
**(cathode)

only takes place under the influence of the discharge from the negative pole.

At this point Crookes introduced a footnote* which presented some
of the conclusions about cathode rays reached by other workers.
These comments not only support the contention that the rays are
emitted from the cathode, but also that objects placed between the
cathode and anode cast a well-defined shadow on the glass behind
the anode. They conclude that the cathode rays travel in straight
lines. After the footnote, Crookes further substantiated this conclusion.

PROJECTION OF MOLECULAR SHADOWS. . . .

The molecular ray which gives birth to green light absolutely
refuses to turn a corner, and radiates from the negative pole in
straight lines, casting strong and sharply-defined shadows of any-
thing which happens to be in its path. In a ∪ tube with poles at each
end, one leg will be bright green and the other almost dark, the light
being cut off sharply by the bend of the glass, a shadow being
projected on the curvature. . . .

Crookes now set up a tube where a star-shaped piece of aluminum
was interposed between the cathode and a piece of uranium glass
which glowed brightly when bombarded by cathode rays.

The whole upper part of the tube which is in the line of direct
projection from the negative pole, glows with an intense yellowish-
green fluorescent light. The uranium plate is still more brilliant, and

*While this paper was passing through the press, my attention has been drawn to two
Memoirs by H. Eugen Goldstein, communicated to the Berlin Academy of Sciences, May 4,
1876, and November 23, 1876, in which some of the results announced in this paper have
been anticipated. . . .
"The negative light which produces this phosphorescence is, as was already assumed by
Hittorf, a rectilinear radiation, which extends from the negative pole into surrounding space.
Still there are essential differences between the diffusion of this remarkable motion and the
likewise rectilinear movement of the light, some of which differences are here brought
forward.
"Hittorf observed that a body placed between the side of the glass and a pointlike cathode,
throws a shadow in the phosphoresent light of the latter.
"Well defined, though not very sharp shadows of small objects may be obtained not merely
from a pointlike or linear negative pole, but also from extended negative surfaces placed at a
small distance from the opaque object.
"A surface which merely radiates light, *e.g.,* an ignited body, under similar conditions
throws a scarcely visible expanded penumbra.
"The negative light is therefore a rectilinear radiation, which is propagated preferably in a
manner almost normal to the producing surface.
"If between the cathode and the green luminous side of the tube there is introduced a solid
body, its shadow is thrown upon the side, since it excludes such rays of the cathode as
impinge upon it from reaching the side. If the solid body after some time is removed, the
shadow disappears, . . .

of a greenish colour. Where the shadow of the star falls on it, no
phosphorescence whatever is visible. . . .

In the next paragraph Crookes concluded from these experiments
that the cathode rays must be particulate and not radiation similar to
light.

> The theory best supported by experiment, and the one which
> although new is not at all improbable in the present state of our
> knowledge respecting molecules, is that the greenish-yellow phos-
> phorescence of the glass is caused by the direct impact of the mole-
> cules on the surface of the glass. The shadows are not optical but
> *molecular* shadows, only they are revealed by an ordinary illuminat-
> ing effect. The sharpness of the shadow, when projected from a wide
> pole, proves them to be molecular. Had the projection from the
> negative pole radiated in all directions, after the manner of light
> radiating from a luminous disk, the shadows would not be perfectly
> sharp, but would be surrounded by a penumbra. Being, however,
> projected material molecules in the same electrical state, they do not
> cross each other, but travel on in slightly divergent paths, giving
> perfectly sharp shadows with no penumbrae. . . .

In the next experiments Crookes demonstrated that the bombardment of cathode rays is capable of causing a paddle wheel or fly to
rotate, indicating that the cathode rays have momentum, and therefore, must consist of particles with mass.

MECHANICAL ACTION OF PROJECTED MOLECULES.

> It was noticed* that when the coil was first turned on, the thin
> glass film was driven back at the moment of becoming phosphores-
> cent. This seemed to point to an actual material blow being given by
> the molecular impact, and the following experiment was devised to
> render this mechanical action more evident.
>
> A large somewhat egg-shaped bulb (Figure 1, elevation) is
> furnished at each end with flat aluminium poles, *a* and *b*; a pointed
> aluminium pole is inserted at *c*. At *d*, a little indicator is suspended
> from jointed glass fibres, so as to admit of being brought into any
> position near the middle of the bulb, by tilting the apparatus. The
> indicator consists of a small radiometer fly 8 millims. in diameter,
> furnished with clear mica vanes 2 millims. across, and delicately
> supported on a glass cup and needle point. A screen, cut out of a flat
> aluminium plate 12 millims. wide and 30 millims. high, is supported
> upright in the bulb at *e*, a little on one side of its axis, being attached
> to the bulb by a platinum wire passing through the glass, so that if
> needed the screen *e* can be used as a pole.
>
> This apparatus was designed with a double object. The indi-
> cator fly is not blacked on one side or favourably presented, there-

*(in a previous experiment)

fore if immersed in a full stream of projected molecules, there will
be no tendency for it to turn one way rather than the other. If,
however, I tilt the bulb so as to bring the indicator half in and half
out of the molecular shadow cast by the screen, I should expect to
see the fly driven round to the right or to the left by the molecules
striking one side only. . . .

The indicator fly is first brought into position . . . where it is
entirely screened from the molecular stream; no movement takes
place. The apparatus is slightly tilted, till the fly comes into position. . . half in and half out of the shadow; very rapid rotation takes
place in the direction of the arrow, showing that impacts occur in
the direction anticipated. The apparatus being further tilted, so as to
bring the indicator quite outside the shadow into position . . . no
movement takes place. When the indicator is brought to the other
side of the shadow into position . . . the rotation is very rapid in the
direction of the arrow, opposite to what it is when at. . . *

It had long been known that a charged particle moving in a straight
line will be deviated from its path by a magnetic field. Hence, if
cathode rays are a stream of charged particles, the presence of a
magnet should cause the cathode rays to be bent away from a
straight line. Crookes now attacked this problem, first with an electromagnet and then with a bar magnet.

MAGNETIC DEFLECTION OF LINES OF MOLECULAR FORCE

An electromagnet is placed beneath the bulb, shown at S in
Figure 1 . . . .

A battery of from 1 to 5 Grove's cells is connected with
the magnet. The current is made to pass in such a direction that the
pole under the bulb . . . is the one which would point towards the
south were the magnet freely suspended. . . .

The electromagnet is now excited by 1 cell. The shadow** is deflected sideways . . .

Figure 1

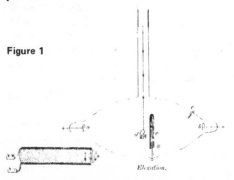

Elevation.

*(the previously partially-shadowed position)
**(cast by an aluminum plate in the cathode ray tube)

On increasing the number of cells actuating the electromagnet, the deflection of the shadow likewise increases. . . .

On reversing the battery current passing round the magnet, the above-named deflections are obtained in the opposite direction. . . .

These phenomena of magnetic deflection are obtained with permanent magnets as well as with electromagnets. . . .

I have spoken of shadows being deflected by the magnet as a convenient way of describing the phenomena observed; but it will be understood that what is really deflected is the path of the molecules driven from the negative pole and whose impact on the phosphorescent surface causes light. The shadows are the effect of a material obstacle in the way of the molecules.

In the apparatus now about to be described, a ray of light was used instead of a shadow. Figure 2 shows the arrangement.

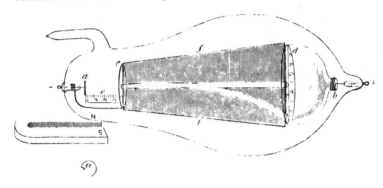

Figure 2.

The poles are at *a* and *b*. The negative pole *a* is a flat aluminium disk with a notch cut in it. The pole *b* is a ring of aluminium; *c* is a mica screen with a small hole in the middle about 1 millim. in diameter; . . . the hole in *c* is to enable a spot of light to be thrown on the scale *d* from the pole *a*; . . . .

the scale on *d* is to enable me to measure the deflection of the ray proceeding from *a*, through *c*, to *d*, when bent by the magnet; *ff* is a vertical screen of mica in the plane of the movement of the ray, covered with a phosphorescent powder. On this the path of the ray traces itself in a straight line when the magnet is absent, and curved when the magnet is present.

The magnet used is a compound horseshoe magnet capable of supporting about 5 lbs. . . .

When two opposite poles, as in a horseshoe magnet, are placed in a line with the normal direction of movement of the molecules, one tends to twist it spirally to the right, and the other to the left . . . The resultant direction is that the ray is bent in a curve, up or down, along a plane at right angles to the plane of the magnet, . . .

The experiments now proceed.

> A spot of green light is now seen projected on the scale
> *d,* bright in the centre and shading off at the edges. The spot is about
> 5.5 millims. diameter. The centre of the spot, the magnet being
> away, is at 18.3 millims. on the scale. On placing the magnet in
> position, the spot of light is bent vertically down to 5 millims. on
> the scale, or a distance of 13.3 millims.
> The magnet would easily bend the light much more if brought
> nearer, . . .

These and similar observations demonstrated clearly the deviation of
the cathode rays in a magnetic field and strongly suggested that
cathode rays are a stream of charged particles.

A last experiment demonstrated that cathode rays could be
focused on a point and that these rays have a great deal of energy. By
using a cup-shaped cathode, Crookes was able to focus cathode rays
on a thin strip of platinum. In his own words:

> Another apparatus was now made, as shown in Figure 3. A
> nearly hemispherical cup of polished aluminium *a* is made one pole
> in a bulb, and a small disk of aluminium *b* is made the other pole. At
> *c* a strip of platinum is held by a wire passing through the glass, and
> forming another pole at *d*. The tip of the platinum strip is brought
> to the centre of curvature, and the whole is exhausted to a very high
> point. On first turning on the induction current, the cup being made
> the negative pole, the platinum strip entered into very rapid vibra-
> tion. This soon stopped, and the platinum quickly rose to a white
> heat, and would have melted had I not stopped the action of the
> coil . . .

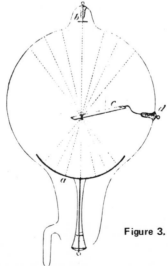

Figure 3.

Crookes now speculated on the results of his experiments:

> I hope I may be allowed to record some theoretical specula-
> tions which have gradually formed in my mind during the progress
> of these experiments. I put them forward only as working hypoth-
> eses, useful, perhaps necessary, in the first dawn of new knowledge,
> but only to be retained as long as they are of assistance; for experi-
> mental research is necessarily and slowly progressive, and one's early
> provisional hypotheses have to be modified, adjusted, perhaps alto-
> gether abandoned in deference to later observations.

AN ULTRA-GASEOUS STATE OF MATTER.

> The modern idea of the gaseous state of matter is based upon
> the supposition that a given space of the capacity of, say, a cubic
> centimetre, contains millions of millions of molecules in rapid mo-
> tion in all directions, each having millions of encounters in a second.
> In such a case the length of the mean free path of the molecules is
> excessively small as compared with the dimensions of the vessel, and
> properties are observed which constitute the ordinary gaseous state
> of matter, and which depend upon constant collisions. But by great
> rarefaction the free path may be made so long that the hits in a given
> time are negligible in comparison to the misses, in which case the
> average molecule is allowed to obey its own motions or laws without
> interference; and if the mean free path is comparable to the dimen-
> sions of the vessel, the properties which constitute gaseity are re-
> duced to a minimum, and the matter becomes exalted to an ultra-
> gaseous or molecular state, in which the very decided but hitherto
> masked properties now under investigation come into play.
>
> The phenomena in these exhausted tubes reveal to physical
> science a new world—a world where matter may exist in a fourth
> state, where the corpuscular theory of light may be true, and where
> light does not always move in straight lines, but where we can never
> enter, and with which we must be content to observe and experi-
> ment from the outside.

It seems clear that Crookes interpreted cathode rays as a beam of
charged atoms or molecules, and considered their characteristics as
resulting from their being in a "fourth *state*" of matter, rather than
from their being a *kind* of matter different from atoms or molecules.

The characteristics of cathode rays, as determined by Crookes,
can be summed up as: (1) cathode rays are composed of charged
particles, (2) they are negatively charged (determined from the fact
that they are repelled by the cathode and determined from the direction of deflection in a magnetic field), (3) they travel in straight
lines, (4) they are high-energy particles.

Strangely enough, in the face of all this evidence, many important workers still considered cathode rays as nonparticulate, but rather as some form of electromagnetic radiation. The major reason for this was that no one had successfully deflected cathode rays with an electric field. Charged particles should be deflected from a straight line path by electric as well as magnetic fields.

It was not until J.J. Thomson's work in the 1890s that the problem of the nature of cathode rays was finally settled. He succeeded in deflecting cathode rays in electric fields and used a combination of electric and magnetic fields to determine the charge to mass ratio of the cathode ray particles. Although not the first to develop these techniques, he was the first to successfully employ them in the laboratory, and to correctly interpret the momentous importance of the results. The major part of his work was presented in an 1897 paper (56), from which the following excerpts are taken.

In the opening paragraph Thomson reviewed the problem and in the second, in a typically scientific manner chose to investigate the theory which offers the best possibility for experimental verification!

> The experiments discussed in this paper were undertaken in the hope of gaining some information as to the nature of the Cathode Rays. The most diverse opinions are held as to these rays; according to the almost unanimous opinion of German physicists they are due to some process in the æther to which—in asmuch as in a uniform magnetic field their course is circular and not rectilinear—no phenomenon hitherto observed is analogous: another view of these rays is that, so far from being wholly ætherial, they are in fact wholly material, and that they mark the paths of particles of matter charged with negative electricity. It would seem at first sight that it ought not to be difficult to discriminate between views so different, yet experience shows that this is not the case, as amongst the physicists who have most deeply studied the subject can be found supporters of either theory.
>
> The electrified-particle theory has for purposes of research a great advantage over the ætherial theory, since it is definite and its consequences can be predicted; with the ætherial theory it is impossible to predict what will happen under any given circumstances, as on this theory we are dealing with hitherto unobserved phenomena in the æther, of whose laws we are ignorant.
>
> The following experiments were made to test some of the consequences of the electrified-particle theory.

Thomson then demonstrated once again that cathode rays are negatively charged particles, and dispelled the last argument against this position.

Charge Carried by the Cathode Rays.

If these rays are negatively electrified particles, then when they enter an enclosure they ought to carry into it a charge of negative electricity. . . .

Perrin found that when the rays passed into the inner cylinder the electroscope received a charge of negative electricity, while no charge went to the electroscope when the rays were deflected by a magnet so as no longer to pass through the hole.

This experiment proves that something charged with negative electricity is shot off from the cathode, travelling at right angles to it, and that this something is deflected by a magnet; . . .

Deflexion of the Cathode Rays by an Electrostatic Field.

An objection very generally urged against the view that the cathode rays are negatively electrified particles, is that hitherto no deflexion of the rays has been observed under a small electrostatic force, . . .

Hertz made the rays travel between two parallel plates of metal placed inside the discharge-tube, but found that they were not deflected when the plates were connected with a battery of storage-cells; on repeating this experiment I at first got the same result, but subsequent experiments showed that the absence of deflexion is due to the conductivity conferred on the rarefied gas by the cathode rays. On measuring this conductivity it was found that it diminished very rapidly as the exhaustion increased; it seemed then that on trying Hertz's experiment at very high exhaustions there might be a chance of detecting the deflexion of the cathode rays by an electrostatic force.

The apparatus used is represented in Figure 4.

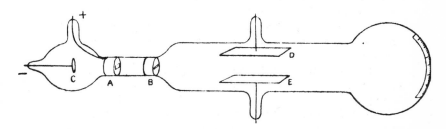

Figure 4.

The rays from the cathode C pass through a slit in the anode A, which is a metal plug fitting tightly into the tube and connected with the earth; after passing through a second slit in another earth-connected metal plug B, they travel between two parallel aluminium plates about 5 cm. long by 2 broad and at a distance of 1·5 cm. apart; they then fall on the end of the tube and produce a narrow well-defined phosphorescent patch. A scale pasted on the outside of the tube serves to measure the deflexion of this patch.

At high exhaustions the rays were deflected when the two aluminium plates were connected with the terminals of a battery of small storage-cells; the rays were depressed when the upper plate was connected with the negative pole of the battery, the lower with the positive, and raised when the upper plate was connected with the positive, the lower with the negative pole. The deflexion was proportional to the difference of potential between the plates, and I could detect the deflexion when the potential-difference was as small as two volts. . . .

A series of measurements of the deflexion of the rays by the electrostatic force under various circumstances will be found later on in the part of the paper which deals with the velocity of the rays and the ratio of the mass of the electrified particles to the charge carried by them. . . .

Having established that cathode rays are negatively charged particles, Thomson now delved deeper into the characteristics of these particles. To do this he determined the ratio of the charge on the cathode particle (e) to its mass (m). The results were of vast importance, since they demolished Dalton's assumption that the atom is indivisible.

As the cathode rays carry a charge of negative electricity, are deflected by an electrostatic force as if they were negatively electrified, and are acted on by a magnetic force in just the way in which this force would act on a negatively electrified body moving along the path of these rays, I can see no escape from the conclusion that they are charges of negative electricity carried by particles of matter. The question next arises, What are these particles? are they atoms, or molecules, or matter in a still finer state of subdivision? To throw some light on this point, I have made a series of measurements of the ratio of the mass of these particles to the charge carried by it. To determine this quantity, I have used two independent methods. . . .

The first method used by Thomson involved measurements with an electroscope of the total charge on a series of cathode ray particles. He also measured kinetic energy of these particles, as indicated by the change in temperature caused by the bombardment of the particles, and the force of a magnetic field deviating the particles from a straight line. The velocity of the particles can also be calculated in these procedures. He made a series of charge to mass ratio (e/m) determinations under different conditions. A summary of his results follows. The tables mentioned are not presented.

> It will be seen from these tables that the value of *m/e* is independent of the nature of the gas. Thus, for the first tube the mean for air is ·40 × 10^{-7}, for hydrogen ·42 × 10^{-7}, and for carbonic acid gas ·4 × 10^{-7}; for the second tube the mean for air is ·52 × 10^{-7}, for hydrogen ·50 × 10^{-7}, and for carbonic acid gas ·54 × 10^{-7}.
>
> Experiments were tried with electrodes made of iron instead of aluminium; this altered the appearance of the discharge and the value of *v* at the same pressure, the values of *m/e* were, however, the same in the two tubes; . . .

The important thing to notice from his results is that the e/m ratio is the same regardless of what gas is in the tube or of what metal the cathode is made.

Thomson then describes his second procedure.

> Before proceeding to discuss the results of these measurements I shall describe another method of measuring the quantities *m/e* and *v* of an entirely different kind from the preceding; this method is based upon the deflexion of the cathode rays in an electrostatic field. If we measure the deflexion experienced by the rays when traversing a given length under a uniform electric intensity, and the deflexion of the rays when they traverse a given distance under a uniform magnetic field, we can find the values of *m/e* and *v* . . .

In the second procedure Thomson applied both an electric and magnetic field to the cathode ray beam, arranged so that the deflection caused by the two fields would be in opposite directions. The fields are adjusted so that the force of each field on the cathde ray particles exactly cancel the other out so that there is no deflection of the particles at all. Since in order to cancel each other out the two forces must be equal, Thomson could set the mathematical description of these forces equal to each other. This allowed him to calculate both the velocity of the particles and their e/m ratio.

A brief description of the experimental set up is followed by the results of a series of e/m determinations, and finally, by his very important interpretation of these results.

> The apparatus used to measure *v* and *m/e* by this means is that represented in Figure 4. The electric field was produced by connect-

ing the two aluminium plates to the terminals of a battery of stor-age-cells. The phosphorescent patch at the end of the tube was de-flected, and the deflexion measured by a scale pasted to the end of the tube. . . .

The magnetic field was produced by placing outside the tube two coils whose diameter was equal to the length of the plates; the coils were placed so that they covered the space occupied by the plates, the distance between the coils was equal to the radius of either. . . .

In the following experiments the current through the coils was adjusted so that the electrostatic deflexion was the same as the magnetic;—

Gas	m/e	v*
Air	1.3×10^{-7}	2.3×10^{9}
Air	1.1×10^{-7}	2.8×10^{9}
Air	1.2×10^{-7}	2.3×10^{9}
Hydrogen	1.5×10^{-7}	2.5×10^{9}
Carbonic Acid	1.5×10^{-7}	2.2×10^{9}
Air	1.3×10^{-7}	3.6×10^{9}
Air	1.1×10^{-7}	2.8×10^{9}

*(cm/sec)

The cathode in the first five experiments was aluminium, in the last two experiments it was made of platinum; . . .

This method of determining the values of m/e and v is much less laborious and probably more accurate than the former method; it cannot, however, be used over so wide a range of pressures.

From these determinations we see that the value of m/e is independent of the nature of the gas, and that its value 10^{-7} is very small compared with the value 10^{-4}, which is the smallest value of this quantity previously known, and which is the value for the hydrogen ion in electrolysis.

Thus for the carriers of the electricity in the cathode rays m/e is very small compared with its value in electrolysis. The smallness of m/e may be due to the smallness of m or the largeness of e, or to a combination of these two. That the carriers of the charges in the cathode rays are small compared with ordinary molecules is shown, I

think, by Lenard's results as to the rate at which the brightness of
the phosphorescence produced by these rays diminishes with the
length of path travelled by the ray. . . .

Later in the paper Thomson stated that he thought that the cathode
ray particle was small and the charge was large, one of his few errors
in this paper. He then continued his interpretations.

> The explanation which seems to me to account in the most
> simple and straightforward manner for the facts is founded on a view
> of the constitution of the chemical elements which has been favour-
> ably entertained by many chemists: this view is that the atoms of
> the different chemical elements are different aggregations of atoms of
> the same kind. In the form in which this hypothesis was enunciated
> by Prout, the atoms of the different elements were hydrogen atoms;
> in this precise form the hypothesis is not tenable, but if we substi-
> tute for hydrogen some unknown primordial substance X, there is
> nothing known which is inconsistent with this hypothesis, . . .

In the following paragraphs Thomson makes a mistake in assuming
that the source of all the cathode ray particles is the gas in the tube.
Actually, most of the particles come from the cathode itself. Thomson used the word "corpuscle" (small body) for the cathode ray
particle. The name electron was applied later by other workers.

> If, in the very intense electric field in the neighbourhood of
> the cathode, the molecules of the gas are dissociated and are split up,
> not into the ordinary chemical atoms, but into these primordial
> atoms, which we shall for brevity call corpuscles; and if these corpus-
> cles are charged with electricity and projected from the cathode by
> the electric field, they would behave exactly like the cathode rays.
> They would evidently give a value of m/e which is independent of
> the nature of the gas and its pressure, for the carriers are the same
> whatever the gas may be; . . .
> Thus on this view we have in the cathode rays matter in a new
> state, a state in which the subdivision of matter is carried very much
> further than in the ordinary gaseous state: a state in which all mat-
> ter—that is, matter derived from different sources such as hydrogen,
> oxygen, &c.—is of one and the same kind; this matter being the
> substance from which all the chemical elements are built up. . . .

Thomson then began to speculate on the arrangement of the cathode
particles in an atom. In succeeding years he developed this speculation into a model of the atom. Already he was trying to explain
Mendeleev's periodic table!

> If we regard the chemical atom as an aggregation of a number
> of primordial atoms, the problem of finding the configurations of
> stable equilibrium for a number of equal particles acting on each
> other according to some law of force . . .

—is of great interest in connexion with the relation between
the properties of an element and its atomic weight. Unfortunately the
equations which determine the stability of such a collection of
particles increase so rapidly in complexity with the number of
particles that a general mathematical investigation is scarcely possible.
We can, however, obtain a good deal of insight into the general laws
which govern such configurations by the use of models, the simplest
of which is the floating magnets of Professor Mayer. In this model the
magnets arrange themselves in equilibrium under their mutual repul-
sions and a central attraction caused by the pole of a large magnet
placed above the floating magnets.

> A study of the forms taken by these magnets seems to me to
> be suggestive in relation to the periodic law.

Thomson had made the big breakthrough. The atom can no
longer be considered a solid sphere. It can no longer be considered
indivisible. At least one particle, the corpuscle (electron) is a part of
the atoms of all elements.

During 1898 and 1899, a great deal was learned about the corpuscle (the word electron had previously been applied to the fundamental unit of charge on an ion, or e, but soon was applied to the
cathode ray particle). Rough determinations of the charge and mass
of the electron were made. If you can determine the e/m ratio, and
can determine either e or m, you can calculate the other. The experimental efforts were directed towards finding e, the charge on the
electron. They involved finding first, the charge on an ion (charged
atom), and then assuming that the charge on the ion was the result of
the ion having one too many or few electrons (compared to its
neutral state).

The first determinations of e were made in 1898 by Townsend
(62). His experimental procedure was based on an observation, described in the next paragraph, that charged particles act as a nucleus
for the condensation of water droplets.

> The experiments which are described in this paper form a
> continuation to those which have already been published in the
> 'Proceedings of the Cambridge Philosophical Society,' vol. ix. pt. v.
> It was there shown that the gases given off by the electrolysis of
> sulphuric acid or caustic potash carry with them an electric charge, a

large percentage of which remains in the gas after it has been bubbled through a liquid, and passed through glass wool to remove the
spray. Another property of these gases is their power of condensing
moisture to form a cloud. No such cloud could be observed in newly
prepared gases unless they were charged, and, further, the weight of
the cloud was found to be proportional to the charge on the gas.
These results go to show that the condensation of the moisture is
connected with the charge; and the experiments described . . .

in Section 19 of this paper prove that when the cloud is
formed in a charged gas the electrification resides on the drops forming the cloud. So that we have definite proof of the fact that the
drops are formed round the carriers of the electric charge.

2. These results were used to find the charge on each carrier, . . .

In his experiments Townsend found the total weight of the water
droplets formed around the charged gas molecules, and with an elec-
trometer, determined the total charge. He then determined the size
of each water droplet by observing how fast the cloud of droplets
settled in air. Since the air retards the fall of little particles more than
big ones, a particle's settling rate in air can be used to measure its
size. From the size of the droplet you can determine its mass.

We thus know the weight of each drop, and dividing this into
the weight of the corresponding cloud, we obtain the number of
drops per cubic centimetre.

When the number of drops is divided into the charge per cubic
centimetre the charge on each is obtained.

The charge on the positive carrier was thus found to be
$2 \cdot 4 \times 10^{-10}$, and that on the negative carrier to be $2 \cdot 9 \times 10^{-10}$.
When we take into account all the experimental errors, these two
charges may be considered equal and approximately 3×10^{-10}.*

J.J. Thomson the same year published his first determination of the
charge on an ion (e) (57). He, like Townsend, determined the size of
the water droplets by observing the rate of fall of the cloud of
droplets under the influence of gravity. He produced the ions by
bombarding the air with X-rays. He employed the marvelous new
"cloud chamber" which had just been constructed by C.T.R. Wilson.
In a cloud chamber a piston suddenly increases the volume of the
chamber. The expansion of the enclosed gas causes it to be cooled.
The air has been saturated with water vapor so that any cooling

*(ESU)

causes the air to become supersaturated. Any charged particle entering the chamber then acts as a nucleus for condensation, and a water
droplet is formed.

> The following experiments were made in order to determine
> the magnitude of the charge of electricity carried by the ions which
> are produced when Röntgen rays pass through a gas.
>
> The theory of the method used is as follows:—By measuring
> the current passing through a gas exposed to Röntgen rays and acted
> upon by a known electromotive force, we determine the value of the
> product *nev*, where *n* is the number of ions in unit volume of the
> gas, *e* the charge on an ion, and *v* the mean velocity of the positive
> and negative ions under the electromotive force to which they are
> exposed.
>
> Mr. Rutherford (Phil. Mag. vol. xliv. p. 422, 1897) has deter-
> mined the value of *v* for a considerable number of gases; using these
> values, the measurement of the current through a gas gives us the
> product *ne*; hence if we can determine *n*, we can deduce the value of
> *e*.
>
> The method I have employed to determine *n* is founded on the
> discovery made by Mr. C.T.R. Wilson (Phil. Trans. A, 1897, p. 265)
> that when Röntgen rays pass through dustfree air a cloud is pro-
> duced by an expansion which is incapable of producing cloudy con-
> densation when the gas is not exposed to these rays. When a determi-
> nate expansion is suddenly produced in dust-free air a definite and
> calculable amount of water is deposited in consequence of the lower-
> ing of the temperature of the air by adiabatic expansion. When the
> gas is exposed to the rays the ions caused by the rays seem to act as
> nuclei around which the water condenses. . . .
>
> If each ion acts as the nucleus for a drop, then if we
> know the size of the drop and the mass of water deposited per unit
> volume, we shall be able to determine the number of drops, and
> hence the number of ions in unit volume of the gas. One part of the
> investigation is thus the determination of the size of the drops: this
> gives us *n;* and as we know from the electrical investigation *ne,* we
> have the means of determining *e*. . . .

Thomson included a series of corrections to improve the reliability of
his results.

> In the preceding investigation we have assumed that the nuclei
> producing the cloud are those which cause the conductivity, and are
> produced by the rays; there is, however, a small cloud produced even
> when no rays are on . . .
>   this makes

$$e = 7{\cdot}4 \times 10^{-10}.$$

The results of other experiments on air are given in the following table:—

	e Uncorrected for Nuclei Present Without Rays	e Corrected
...	6.7×10^{-10} 6.4 7.3 6.3 5.0	7.6 7.2 8.4 7.4 6.0

The mean of these values and the one previously obtained is

$e = 7 \cdot 3 \times 10^{-10}$ electrostatic units.

Another correction has to be made to allow for the conductivity of the walls of the vessel A due to the film of moisture with which it is coated. . . .

Applying this correction the mean value of e is equal to

$$\frac{8}{9} \times 7 \cdot 3 \times 10^{-10} = 6 \cdot 5 \times 10^{-10}. \dots$$

He then made a series of experiments using hydrogen instead of air.

Mean	6.7×10^{-10}

The value of e for hydrogen has not been corrected in the way that the value of e for air has been by allowing for the part of the cloud formed independently of the rays. Allowing for this the experiments seem to show that the charge on the ion in hydrogen is the same as in air.

Thomson did not speculate at any length about the significance of his findings or their connection with cathode ray particles, but did so at length in the next paper presented (58). In this paper he put everything together. The existence of a subatomic particle, common to the atoms of all elements is here more firmly established.

In a former paper (Phil. Mag. Oct. 1897) I gave a determination of the value of the ratio of the mass, m, of the ion to its charge, e, in the case of the stream of negative electrification which constitutes the cathode rays. . . .

In these experiments it was only the value of m/e which was determined, and not the values of m and e separately. It was thus possible that the smallness of the ratio might be due to e being greater than the value of the charge carried by the ion in electrolysis rather than to the mass m being very much smaller. Though there were reasons for thinking that the charge e was not greatly different from the electrolytic one, and that we had here to deal with masses smaller than the atom, yet, as these reasons were somewhat indirect, I desired if possible to get a direct measurement of either m or e as well as of m/e. In the case of cathode rays I did not see my way to do this; but another case, where negative electricity is carried by charged particles (*i.e.* when a negatively electrified metal plate in a gas at low pressure is illuminated by ultraviolet light), seemed more hopeful, as in this case we can determine the value of e by the method I previously employed to determine the value of the charge carried by the ions produced by Röntgen-ray radiation (Phil. Mag. Dec. 1898). The following paper contains an account of measurements of m/e and e for the negative electrification discharged by ultraviolet light, and also of m/e for the negative electrification produced by an incandescent carbon filament in an atmosphere of hydrogen. I may be allowed to anticipate the description of these experiments by saying that they lead to the result that the value of m/e in the case of the ultraviolet light, and also in that of the carbon filament, is the same as for the cathode rays; and that in the case of the ultraviolet light, e is the same in magnitude as the charge carried by the hydrogen atom in the electrolysis of solutions. In this case, therefore, we have clear proof that the ions have a very much smaller mass than ordinary atoms; so that in the convection of negative electricity at low pressures we have something smaller even than the atom, something which involves the splitting up of the atom, inasmuch as we have taken from it a part, though only a small one, of its mass. . . .

The following table gives the result of some experiments; . . .

	$e \times 10^{10}$
. . .	7.9
	7.3
	5.3
	7.3
	6
	7

The mean value of e is $6 \cdot 8 \times 10^{-10}$. The values differ a good deal, but we could not expect a very close agreement unless we could procure an absolutely constant source of ultraviolet light, as these experiments are very dependent on the constancy of the light: . . .

The value of e found by me previously for the ions produced by Röntgen rays was $6 \cdot 5 \times 10^{-8}$: hence we conclude that e for the ions produced by ultraviolet light is the same as e for the ions produced by the Röntgen rays; and as Mr. Townsend has shown that the charge on these latter ions is the same as the charge on an atom of hydrogen in electrolysis, we arrive at the result previously referred to, that the charge on the ion produced by ultraviolet light is the same as that on the hydrogen ion in ordinary electrolysis.

The experiments just described, taken in conjunction with previous ones on the value of m/e for the cathode rays (J.J. Thomson, Phil. Mag. Oct. 1897), show that in gases at low pressures negative electrification, though it may be produced by very different means, is made up of units each having a charge of electricity of a definite size; the magnitude of this negative charge is about 6×10^{-10} electrostatic units, and is equal to the positive charge carried by the hydrogen atom in the electrolysis of solutions.

In the latter part of this paper Thomson began to construct his model of the atom which we shall return to later in this chapter. First, however, we should look at a series of experiments performed during the first dozen or so years of the 20th century, in which more and more refined techniques were employed to determine e with greater accuracy. The significance of an accurate determination of e is discussed in several of these papers.

In the first of these papers Wilson (66) uses an experimental procedure similar to Thomson's. However, in addition to recording the settling velocity of the cloud in air, he also introduced an electric field part of the time which exerted a force on the cloud in the same direction as gravity, and hence, increased the cloud's settling velocity. After measuring the two velocities, he could calculate e.

The experiments described in this paper were undertaken with the object of making a fresh determination of the charge on one ion. This charge will throughout this paper be denoted by e.

Prof. Townsend (Phil. Mag. Feb. 1898), in a paper on the "Electrical Properties of Newly Prepared Gases," has described a determination of the average charge on the droplets composing the cloud formed when newly prepared oxygen is bubbled through water. This charge was found to be about 3×10^{-10} electrostatic units of electricity. There are some reasons for supposing that each droplet contains one ion, and consequently Townsend's result may be regarded as a determination of the charge on one ion. The result which I have obtained is in very good agreement with his.

Prof. J.J. Thomson (Phil. Mag. Dec. 1898 and 1899) has given two estimates of e, the first depending on a determination of the average charge on the droplets of a cloud formed by condensation of water-vapour on the ions produced in air by Röntgen rays, and the second on a similar determination for the ions given off by a zinc plate under the action of ultraviolet light. The mean result of the first research was $e = 6 \cdot 5 \times 10^{-10}$ and of the second $e = 6 \cdot 8 \times 10^{-10}$.

Since from the value of e the number of molecules in a cubic centimetre of a gas can be immediately deduced, and also since the absolute value of e is of considerable interest in itself, a fresh determination by a different method appeared to be worth making.

The method I have used depends, like Prof. Thomson's, on the fact discovered by C.T.R. Wilson that the ions produced in air by Röntgen rays act as nuclei for the cloudy condensation of water-vapour when supersaturation exceeding a definite amount is produced by a sudden expansion.

The droplets of the cloud produced presumably each contain one or more ions. Let a droplet containing one ion, and consequently having a charge e, have a mass m which can be determined by observing its rate of fall (r_1 say) in air. If now a vertical electrostatic field of strength X is applied to this droplet there will be a vertical force on the droplet equal to Xe due to the field, so that the total force on the droplet will be $Xe + mg$ where g is the acceleration due to gravity, and reckoning Xe positive when it is in the same direction as the weight mg. Now the rate of steady motion of a sphere in a viscous fluid is proportional to the force acting on it, so that the rate of fall of the droplet will be altered by the electric field. . . .

The principal advantages of my method are that it is not necessary to estimate either the number of drops in the cloud, or the number of ions present at the moment of its formation, or to make the assumption that each droplet contains only one ion. Both these estimations involve assumptions which in practice can only be approximately true, and there is always a danger that some of the drops in the cloud contain more than one ion. . . .

The mean result of the present experiments, viz. $e = 3 \cdot 1 \times 10^{-10}$ of an electrostatic unit, cannot be very far from the truth. I think that it may be considered established by these experiments that e lies between 2×10^{-10} and 4×10^{-10} E.S. unit.

The increasing refinement of method culminated with Millikan and his famous "oil drop" experiment. In the first paper quoted below (25), his chief refinement consisted of making the determinations of the charge on a *single* droplet of water, rather than the entire cloud. He began the paper by stressing the importance of finding e.

> Among all physical constants there are two which will be universally admitted to be of predominant importance; the one is the velocity of light, which now appears in many of the fundamental equations of theoretical physics, and the other is the ultimate, or elementary, electrical charge, a knowledge of which makes possible a determination of the absolute values of all atomic and molecular weights, the absolute number of molecules in a given weight of any substance, the kinetic energy of agitation of any molecule at a given temperature, and a considerable number of other important physical quantities.
>   While the velocity of light is now known with a precision of one part in twenty thousand, the value of the elementary electrical charge has until very recently been exceedingly uncertain. The results herewith presented seem to show that the method here used for its determination—a modification of the Thomson-Wilson cloud method—furnishes the value of *e* with a directness, certainty, and precision, easily comparable with that obtained by any of the methods which have thus far been used, the error in the final mean value being not more than 2 per cent. . . .

At this point Millikan discussed previous determinations of e, and pointed out their various sources of possible error, with emphasis on error introduced by the evaporation of the water droplets. He then demonstrated how he had limited this last source of error.

> My original plan for eliminating the evaporation error was to obtain, if possible, an electric field strong enough to exactly balance the force of gravity upon the cloud and by means of a sliding contact to vary the strength of this field so as to hold the cloud balanced throughout its entire life. In this way it was thought that the whole evaporation history of the cloud might be recorded, and suitable allowances then made in the observations on the rate of fall to eliminate entirely the error due to evaporation. It was not found possible to balance the cloud as had been originally planned, but it was found possible to do something very much better: namely, to hold individual charged drops suspended by the field for periods varying from 30 to 60 seconds. I have never actually timed drops which lasted more than 45 seconds, although I have several times observed drops which in my judgment lasted considerably longer

than this. The drops which it was found possible to balance by an electrical field always carried multiple charges, and the difficulty experienced in balancing such drops was less than had been anticipated.

The procedure is simply to form a cloud and throw on the field immediately thereafter. The drops which have charges of the same sign as that of the upper plate or two weak charges of the opposite sign, rapidly fall, while those which are charged with too many multiples of sign opposite to that of the upper plate are jerked up against gravity to this plate. The result is that after a lapse of 7 or 8 seconds the field of view has become quite clear save for a relatively small number of drops which have just the right ratio of charge to mass to be held suspended by the electric field. These appear as perfectly distinct bright points. I have on several occasions obtained but one single such "star" in the whole field and held it there for nearly a minute. . . .

The life history of these drops is as follows. If they are a little too heavy to be held quite stationary by the field they begin to move slowly down under gravity. Since, however, they slowly evaporate, their downward motion presently ceases, and they become stationary for a considerable period of time; then the field gets the better of gravity and they move slowly upward. . . .

I have often held drops through a period of from 10 to 15 seconds, during which it was impossible to see that they were moving at all. . . .

All the charged drops Millikan viewed had multiple charges (a charge representing more than a single electron). In order to determine the charge on a single electron, it was necessary for him to find the largest charge which would divide evenly into *all* charges on the many drops he investigated.

It will be observed that the only possible elementary charge of which the observed charges are multiples is $4 \cdot 65 \times 10^{-10}$, and further that the measured charges represent all the possible multiples of this charge between 2 and 6. This shows that the elementary charge cannot possibly be the smallest charge which we observed: namely, $9 \cdot 3 \times 10^{-10}$, since we obtained odd as well as even multiples of one half of this quantity.

The advantages of this modified method for the determination of e were summarized by him in an earlier paper (24).

This modification of the cloud method of determining *e* consists:

(1) In making observations, not upon the surface of a cloud, but upon single isolated drops carrying multiple charges.

(2) In exactly balancing gravity upon these single charged drops by an electrical field.

(3) In observing the rate of fall of these same drops under gravity after the electrical field has been thrown off.

(4) In eliminating any possible error due to evaporation by first obtaining stationary, *i.e.*, balanced, drops, and then measuring the times of passage of these same drops across equal spaces in the field of the reading telescope.

(5) In directly measuring the temperature of the cloud chamber instead of computing it. (It is this computation which was found to have introduced the chief error into preceding determinations of *e* by the cloud method. ...

The method compares favorably in directness and precision with any which has thus far been used for determining *e*. The results of 7 very concordant observations on water drops carrying triple positive charges gave $e = 4 \cdot 59 \times 10^{-10}$: of 11 concordant observations on alcohol drops carrying double positive charges gave $e = 4 \cdot 64 \times 10^{-10}$: of 10 observations on water drops carrying quadruple positive charges gave $e = 4 \cdot 56 \times 10^{-10}$: of 5 observations on water drops carrying quintuple positive charges gave $e = 4 \cdot 83 \times 10^{-10}$: of three observations on water drops carrying sextuple positive charges gave $e = 4 \cdot 69 \times 10^{-10}$: of two observations on water drops carrying double positive charges gave $e = 4 \cdot 87 \times 10^{-10}$. The weighted mean of these results is $4 \cdot 65 \times 10^{-10}$. The error in this determination is estimated as not more than 2 per cent.

In the famous "oil drop" experiment Millikan reduced still further possible error by using minute oil drops in place of the water drops. The oil drops did not evaporate to any significant degree while they were being observed. Since this is the culminating and most famous of this series of experiments, it is quoted in some detail (26).

In a preceding paper a method of measuring the elementary electrical charge was presented which differed essentially from methods which had been used by earlier observers only in that all of the measurements from which the charge was deduced were made upon one individual charged carrier. This modification eliminated the chief sources of uncertainty which inhered in preceding determinations by similar methods ...

The sources of error or uncertainty which still inhered in the method arose from: (1) the lack of complete stagnancy in the air through which the drop moved; (2) the lack of perfect uniformity in the electrical field used; (3) the gradual evaporation of the drops, rendering it impossible to hold a given drop under observation for more than a minute, or to time the drop as it fell under gravity alone through a period of more than five or six seconds; (4) the assumption of the exact validity of Stokes's law* for the drops used. The present modification of the method is not only entirely free from all of these limitations, but it constitutes an entirely new way of studying ionization and one which seems to be capable of yielding important results in a considerable number of directions.

With its aid it has already been found possible:

(1) To catch upon a minute droplet of oil and to hold under observation for an indefinite length of time one single atmospheric ion or any desired number of such ions between 1 and 150.

(2) To present direct and tangible demonstration through the study of the behavior in electrical and gravitational fields of this oil drop carrying its captured ions, of the correctness of the view advanced many years ago and supported by evidence from many sources that all electrical charges, however produced, are exact multiples of one definite, elementary, electrical charge, or in other words, that an electrical charge instead of being spread uniformly over the charged surface has a definite granular structure, consisting, in fact, of an exact number of specks, or atoms of electricity, all precisely alike, peppered over the surface of the charged body.

(3) To make an exact determination of the value of the elementary electrical charge which is free from all questionable theoretical assumptions and is limited in accuracy only by that attainable in the measurement of the coefficient of viscosity of air. . . .

(5) To demonstrate that the great majority, if not all, of the ions of ionized air, of both positive and negative sign, carry the elementary electrical charge. . . .

2. THE METHOD

The only essential modification in the method consists in replacing the droplet of water or alcohol by one of oil, mercury or some other nonvolatile substance, and in introducing it into the observing space in a new way.

Figure 5 shows the apparatus used in the following experiments. By means of a commercial "atomizer" A a cloud of fine droplets of oil is blown with the aid of dust-free air into the dust-free chamber C. One or more of the droplets of this cloud is allowed to fall through a pinhole p into the space between the plates M,N of a horizontal air condenser and the pinhole is then closed by means of an electromagnetically operated cover not shown in the diagram. If the pinhole is left open air currents are likely to pass through it and produce irregularities. The plates M,N are heavy, circular ribbed brass castings 22 cm in diameter having surfaces which are ground so

---

*(determining the size of a particle from its rate of fall in air)

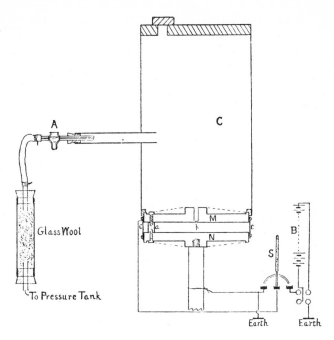

Figure 5.

nearly to true planes that the error is nowhere more than .02 mm.
These planes are held exactly 16 mm apart by means of three small
ebonite posts, held firmly in place by ebonite screws. A strip of thin-
sheet ebonite *C* passes entirely around the plates, thus forming a completely enclosed air space. Three glass windows, 1.5 cm square, are
placed in this ebonite strip at the angular positions $0°$, $165°$, and
$180°$. A narrow parallel beam of light from an arc lamp enters the
condenser through the first window and emerges through the last.

      The other window serves for observing, with the aid of a
short focus telescope placed about 2 feet distant, the illuminated oil
droplet as it floats in the air between the plates. The appearance of
this drop is that of a brilliant star on a black background. It falls, of
course under the action of gravity, toward the lower plate; but before it reaches it, an electrical field of strength between 3,000 volts
and 8,000 volts per centimeter is created between the plates by
means of the battery *B*, and, if the droplet had received a frictional
charge of the proper sign and strength as it was blown out through
the atomizer, it is pulled up by this field against gravity, toward the
upper plate. Before it strikes it the plates are short-circuited by
means of the switch *S* and the time required by the drop to fall
under gravity the distance corresponding to the space between the
cross hairs of the observing telescope is accurately determined. Then
the rate at which the droplet moves up under the influence of the

field is measured by timing it through the same distance when the field is on. This operation is repeated and the speeds checked an indefinite number of times, or until the droplet catches an ion from among those which exist normally in air, or which have been produced in the space between the plates by any of the usual ionizing agents like radium or X-rays. The fact that an ion has been caught and the exact instant at which the event happened is signaled to the observer by the change in the speed of the droplet under the influence of the field. From the sign and magnitude of this change in speed, taken in connection with the constant speed under gravity, the sign and the exact value of the charge carried by the captured ion are determined. The error in a single observation need not exceed one third of one percent. It is from the values of the speeds observed that all of the conclusions above mentioned are directly and simply deduced.

The experiment is particularly striking when, as often happens, the droplet carries but one elementary charge and then by the capture of an ion of opposite sign is completely neutralized so that its speed is altogether unaffected by the field. In this case the computed charge is itself the charge on the captured ion. . . .

Table I presents the record of the observations taken upon a drop which was watched through a period of four and one half hours as it was alternately moved up and down between the cross hairs of the observing telescope under the influence of the field F and gravity G. How completely the errors arising from evaporation, convection currents, or any sort of disturbances in the air were eliminated is shown by the constancy during all this time in the value of the velocity under gravity. . . .

For brevity, only the first few values in the two page table are presented here. n is the number of extra electrons on each oildrop. e_n is the total charge on the oil drop, e_1 is charge on a single electron. The mean value is for all the e_1 presented in the original table.

TABLE I.

n	$e_n \times 10^{10}$	$e_1 \times 10^{10}$
7	34.47	4.923
8	39.45	4.931
9	44.42	4.936
10	49.41	4.941
8	39.45	
12	59.12	4.927
9	44.42	
10	49.41	
11	53.92	4.902
	Mean of all e_1's = 4.917	

Now it will be seen from the first observation given in the table that the charge which was originally upon this drop and which was obtained, not from the ions in the air, but from the frictional process involved in blowing the spray, was $34 \cdot 47 \times 10^{-10}$. This number comes within one seventh of one per cent of being exactly seven times the charge on the positive, or on the negative, ion caught in the observations under consideration. In the interval between December, 1909, and May, 1910, Mr Harvey Fletcher and myself took observations in this way upon hundreds of drops which had initial charges varying between the limits 1 and 150, and which were upon as diverse substances as oil, mercury and glycerine and found in every case the original charge on the drop an exact multiple of the smallest charge which we found that the drop caught from the air. The total number of changes which we have observed would be between one and two thousand, and *in not one single instance has there been any change which did not represent the advent upon the drop of one definite invariable quantity of electricity, or a very small multiple of that quantity.* These observations are the justification for assertions 1 and 2 of the introduction.

For the sake of exhibiting in another way the multiple relationship shown by the charges on a given drop the data of Table I. have been rearranged in the form shown in Table II.

In Table II, the observed charge on the drops was compared to the value predicted on the basis of e being 4.917×10^{-10} ESU (the mean results from this experiment). The closeness of the compared values indicated the degree of accuracy achieved with this experimental procedure.

TABLE II.

n	4.917 x n	Observed Charge	n	4.917 x n	Observed Charge
1	4.917		10	49.17	49.41
2	9.834		11	54.09	53.92
3	14.72		12	59.00	59.12
4	19.66	19.66	13	63.92	63.68
5	24.59	24.60	14	68.84	68.65
6	29.50	29.62	15	73.75	
7	34.42	34.47	16	78.67	78.34
8	39.34	39.38	17	83.59	83.22
9	44.25	44.42	18	88.51	

No more exact or more consistent multiple relationship is found in the data which the chemists have amassed on combining powers, and upon which the atomic theory of matter rests, than is found in tables I to XIII.....

Later in the paper Millikan, by discovering a computation error and employing slightly improved methods of calculation, altered the mean value for e to that listed below. Included with e are several other important quantities which can be calculated once you have determined e.

> In conclusion there is presented a summary of the most important of the molecular magnitudes, accurate values of which are made possible by an accurate determination of *e*. . . .
>
> e = 4.891 $\times$ 10^{-10} E.S.U. the smallest quantity of electricity capable of separate existence. *
>
> N = 5.922 $\times$ 10^{23} the number of molecules in one gram molecule of any substance.
>
> n = 2.644 $\times$ 10^{19} the number of molecules in 1 c.c. of any gas at $0°$C. and 76 cm.
>
> . . .
>
> m = 1.702 $\times$ 10^{-24} gms. the weight of the hydrogen atom.

Slight alterations in Millikan's results were necessitated years later by new information, but the "oil drop" experiment stands as one of the great experimental procedures in the history of science.

Let us return now to 1898 and see how J.J. Thomson and others used the discovery of the electron as the basis for a model of the atom quite different from the indivisible, solid sphere model of Dalton.

We have previously mentioned that Thomson's 1899 paper (58) contained not only a determination of the value of e, but also contained his speculation concerning a model of the atom. Taking up where we left off on page 25, we can follow through his reasoning.

> In gases at low pressures these units of negative electric charge are always associated with carriers of a definite mass. This mass is exceedingly small, being only about 1·4 $\times$ 10^{-3} of that of the hydrogen ion, the smallest mass hitherto recognized as capable of a separate existence. The production of negative electrification thus involves the splitting up of an atom, as from a collection of atoms something is detached whose mass is less than that of a single atom. . . .

In the following paragraph Thomson stated that although there is a fundamental negatively charged particle common to the atoms of all

*(charge on an electron)

elements (the corpuscle or electron), there is no evidence that a fundamental positively charged particle exists.

Hitherto we have been considering only negative electrification; as far as our present knowledge extends positive electrification is never associated with masses as small as those which invariably accompany negative electrification in gases at low pressures. From W. Wien's experiments on the ratio of the mass to the electric charge for the carriers of positive electrification in a higly exhausted vacuum-tube . . . ,it would seem that the masses with which positive electrification is associated are comparable with the masses of ordinary atoms. .. . .

There are some other phenomena which seem to have a very direct bearing on the nature of the process of ionizing a gas. Thus I have shown (Phil. Mag. Dec. 1898) that when a gas is ionized by Röntgen rays, the charges on the ions are the same whatever the nature of the gas: thus we get the same charges on the ions whether we ionize hydrogen or oxygen. . . .

These results, taken in conjunction with the measurements of the mass of the negative ion, suggest that the ionization of a gas consists in the detachment from the atom of a negative ion; this negative ion being the same for all gases, while the mass of the ion is only a small fraction of the mass of an atom of hydrogen.

From what we have seen, this negative ion must be a quantity of fundamental importance in any theory of electrical action; indeed, it seems not improbable that it is the fundamental quantity in terms of which all electrical processes can be expressed. For, as we have seen, its mass and its charge are invariable, independent both of the processes by which the electrification is produced and of the gas from which the ions are set free. It thus possesses the characteristics of being a fundamental conception in electricity; and it seems desirable to adopt some view of electrical action which brings this conception into prominence. These considerations have led me to take as a working hypothesis the following method of regarding the electrification of a gas, or indeed of matter in any state.

I regard the atom as containing a large number of smaller bodies which I will call corpuscles; these corpuscles are equal to each other; the mass of a corpuscle is the mass of the negative ion in a gas at low pressure, *i.e.* about 3×10^{-26} of a gramme. In the normal atom, this assemblage of corpuscles forms a system which is electrically neutral. Though the individual corpuscles behave like negative ions, yet when they are assembled in a neutral atom the negative effect is balanced by something which causes the space through which the corpuscles are spread to act as if it had a charge of positive electricity equal in amount to the sum of the negative charges on the corpuscles. Electrification of a gas I regard as due to the splitting up of some of the atoms of the gas, resulting in the detachment of a corpuscle from some of the atoms. The detached corpuscles behave like negative ions, each carrying a constant negative charge, which we shall call for brevity the unit charge; while the part of the atom left behind behaves like a positive ion with the unit positive charge and a mass large compared with that of the negative ion. On this view, electrification essentially involves the splitting up of the atom, a part of the mass of the atom getting free and becoming detached from the original atom.

A positively electrified atom is an atom which has lost some of its "free mass," and this free mass is to be found along with the corresponding negative charge. Changes in the electrical charge on an atom are due to corpuscles moving from the atom when the positive charge is increased, or to corpuscles moving up to it when the negative charge is increased. Thus when anions and cations are liberated against the electrodes in the electrolysis of solutions, the ion with the positive charge is neutralized by a corpuscle moving from the electrode to the ion, while the ion with the negative charge is neutralized by a corpuscle passing from the ion to the electrode. The corpuscles are the vehicles by which electricity is carried from one atom to another.

We are thus led to the conclusion that the mass of an atom is not invariable: that, for example, if in the molecule of HCl the hydrogen atom has the positive and the chlorine atom the negative charge, then the mass of the hydrogen atom is less than half the mass of the hydrogen molecule H_2; while, on the other hand, the mass of the chlorine atom in the molecule of HCl is greater than half the mass of the chlorine molecule Cl_2.

The amount by which the mass of an atom may vary is proportional to the charge of electricity it can receive; and as we have no evidence that an atom can receive a greater charge than that of its ion in the electrolysis of solutions, and as this charge is equal to the valency of the ion multiplied by the charge on the hydrogen atom, we conclude that the variability of the mass of an atom which can be produced by known processes is proportional to the valency of the atom, and our determination of the mass of the corpuscle shows that this variability is only a small fraction of the mass of the original atom.

Thomson then suggested that there are a considerable number of corpuscles (electrons) in an atom, but didn't speculate as to their possible arrangement within the atom.

In 1902, Lord Kelvin published a paper in which he speculated about the nature of the atom. The model is very much along the lines of Thomson's but seems to be more theoretically based. He made no reference to any of Thomson's work, and seemed to rely little on the physical evidence accumulated by workers with cathode ray tubes (21).

The Aepinus' fluid Kelvin mentions in the first paragraph refers to the hypothetical electrical fluid of the one fluid theory of electricity. A footnote refers back to an earlier paper he had published in 1897 which includes some history of the terminology of the electron.

> My suggestion is that the Aepinus' fluid consists of exceedingly minute equal and similar atoms, which I call electrions†, much smaller than the atoms of ponderable matter; and that they permeate freely through the spaces occupied by these greater atoms and also freely through space not occupied by them. . . .
>
> †Indeed, Faraday's law of electrochemical equivalence seems to necessitate something atomic in electricity, and to justify Johnstone Stoney's word *electron*. The older, and at present even more popular, name *ion* given sixty years ago by Faraday, suggests a convenient modification of it, *electrion,* to denote an atom of resinous electricity. . . .
>
> I shall therefore assume that our electrions act as extremely minute particles of *resinously** electrified matter; that a void atom acts simply as a little globe of atomic substance, possessing as an essential quality vitreous** electricity uniformly distributed through it or through a smaller concentric globe; and that ordinary ponderable matter, not electrified, consists of a vast assemblage of atoms, not void, but having within the portions of space which they occupy just enough of electrions to annul electric force.

This was essentially the model Thomson defended two years later with a more quantitive approach (59). This model has become known as the "plum pudding" model of the atom, since the electrons, imbedded in a sphere of positive charge, would resemble plums in a pudding.

In this paper he attacked the question which any model of the atom must be able to answer: how does the model account for the periodic chemical properties of the elements? His solution, involving rings of electrons is very ingenious. It suffers however, from the fact that there was no way at the time to determine how many electrons were associated with a neutral atom of any element.

*(negative)
**(positive)

The view that the atoms of the elements consist of a number of negatively electrified corpuscles enclosed in a sphere of uniform positive electrification suggests, among other interesting mathematical problems, the one discussed in this paper, that of the motion of a ring of n negatively electrified particles placed inside a uniformly electrified sphere. . . .

We suppose that the atom consists of a number of corpuscles moving about in a sphere of uniform positive electrification: the problems we have to solve are (1) what would be the structure of such an atom, *i.e.,* how would the corpuscles arrange themselves in the sphere; and (2) what properties would this structure confer upon the atom. The solution of (1) when the corpuscles are constrained to move in one plane is indicated by the results we have just obtained— the corpuscles will arrange themselves in a series of concentric rings. This arrangement is necessitated by the fact that a large number of corpuscles cannot be in stable equilibrium when arranged as a single ring, while this ring can be made stable by placing inside it an appropriate number of corpuscles. When the corpuscles are not constrained to one plane, but can move about in all directions, they will arrange themselves in a series of concentric shells; for we can easily see that, as in the case of the ring, a number of corpuscles distributed over the surface of a shell will not be in stable equilibrium if the number of corpuscles is large, unless there are other corpuscles inside the shell, while the equilibrium can be made stable by introducing within the shell an appropriate number of other corpuscles.

The analytical and geometrical difficulties of the problem of the distribution of the corpuscles when they are arranged in shells are much greater than when they are arranged in rings, and I have not as yet succeeded in getting a general solution. We can see, however, that the same kind of properties will be associated with the shells as with the rings; and as our solution of the latter case enables us to give definite results, I shall confine myself to this case, and endeavour to show that the properties conferred on the atom by this ring structure are analogous in many respects to those possessed by the atoms of the chemical elements, and that in particular the properties of the atom will depend upon its atomic weight in a way very analogous to that expressed by the periodic law. . . .

In the next paragraph we see that Thomson thought that the mass of an atom was equal to the total mass of all its electrons. This would necessitate very large numbers of electrons in an atom as the mass of the electron is only about 1/2000 the mass of the hydrogen atom. This did allow him to relate erroneously the periodicity of the elements with atomic weight. Remember that "corpuscle" is Thomson's word for electron.

Let us now consider the connexion between these results and the properties possessed by the atoms of the chemical elements. We suppose that the mass of an atom is the sum of the masses of the corpuscles it contains, so that the atomic weight of an element is measured by the number of corpuscles in its atom. An inspection of the results just given will show that systems built up of rings of corpuscles in the way we have described, will possess properties analogous to some of those possessed by the atom. In the first place, we see that the various arrangements of the corpuscles can be classified in families, the grouping of the corpuscles in the various members of the family having certain features in common. . . .

He then attempted to explain the formation of ions.

Thus when this ring is subjected to disturbances from an external source, one or more corpuscles may easily be detached from it; such an atom therefore will easily lose a negatively electrified corpuscle, and thus acquire a charge of positive electricity; such an atom would behave like the atom of a strongly electropositive element. . . .

When the stability of the outer ring gets very great, it may be possible for one or more corpuscles to be on the surface of the atom without breaking up the ring; in this case the atom could receive a charge of negative electricity, and would behave like the atom of an electronegative element.

Although, as we shall see in later chapters, Thomson's model of the atom was woefully inadequate, it was the first real attempt to replace Dalton's model of the atom with a model based on new experimental evidence. It was correct in the assumption that the atom is divisible, and that the explanation of the periodicity of the chemical properties of the elements lies in the arrangement of subatomic particles.

Chapter 3

Radioactivity

As has often been the case, further major advances in our knowledge of the atom followed only after scientists took an entirely new approach to the study of the nature of matter. This new approach resulted from two accidental discoveries made in 1895 and 1896. The first of these was Roentgen's discovery of X-rays, and the second Becquerel's discovery of radioactivity. Though both of these discoveries are interesting in themselves, it was the brilliant follow up of these discoveries by Rutherford and others, that culminated in the nuclear model of the atom. This is essentially the model of the atom which is still accepted today. In this chapter we will examine the original discoveries and follow through the unraveling of the complex nature of radioactivity.

Roentgen's discovery of X-rays was quite accidental, but his follow up of the discovery was very systematic and thorough. He had been working with a cathode ray tube enclosed in paper in a darkened room. There happened to be a piece of paper, coated with a fluorescent salt nearby, which surprisingly glowed when the cathode ray tube was turned on. It had been long established that cathode rays could pass through only a few centimeters of air, and thus, they could not account for the fluorescence of the paper. Roentgen reports his investigations of the problem in the following paragraphs (30).

ON A NEW KIND OF RAYS

A discharge from a large induction coil is passed through a Hittorf's vacuum tube, or through a well-exhausted Crookes' or Lenard's tube. The tube is surrounded by a fairly close-fitting shield of black paper: it is then possible to see, in a completely darkened room, that paper covered on one side with barium platinocyanide lights up with brilliant fluorescence when brought into the neighbourhood of the tube, whether the painted side or the other be turned towards the tube. The fluorescence is still visible at two metres distance. It is easy to show that the origin of the fluorescence lies within the vacuum tube.

It is seen, therefore, that some agent is capable of penetrating black cardboard which is quite opaque to ultra-violet light, sunlight, or arc-light. It is therefore of interest to investigate how far other bodies can be penetrated by the same agent. It is readily shown that all bodies possess this same transparency, but in very varying degrees. For example, paper is very transparent; the fluorescent screen will light up when placed behind a book of a thousand pages; printer's ink offers no marked resistance. Similarly the fluorescence shows behind two packs of cards; a single card does not visibly diminish the brilliancy of the light. So, again, a single thickness of tinfoil hardly casts a shadow on the screen; several have to be superposed to produce a marked effect. Thick blocks of wood are still transparent. Boards of pine two or three centimetres thick absorb only very little. A piece of sheet aluminium, 15 mm. thick, still allowed the X-rays (as I will call the rays, for the sake of brevity) to pass, but greatly reduced the fluorescence. Glass plates of similar thickness behave similarly; lead glass is, however, much more opaque than glass free from lead. Ebonite several centimeters thick is transparent. If the hand be held before the fluorescent screen, the shadow shows the bones dark with only faint outlines of the surrounding tissues. . . .

The fluorescence of barium platinocyanide is not the only noticeable action of the X-rays. It is to be observed that other bodies exhibit fluorescence, *e.g.*, calcium sulphide, uranium glass, Iceland spar, rock-salt, &c. . . .

The photographic plate can be exposed to the action without removal of the shutter of the dark slide or other protecting case, so that the experiment need not be conducted in darkness. Manifestly, unexposed plates must not be left in their box near the vacuum tube.

It seems now questionable whether the impression on the plate is a direct effect of the X-rays, or a secondary result induced by the fluorescence of the material of the plate. Films can receive the impression as well as ordinary dry plates. . . .

Roentgen's mention of fluorescence was to lead, indirectly, to the discovery of radioactivity. Roentgen ended his report with his analysis of the nature of the rays.

A further distinction, and a noteworthy one, results from the action of a magnet. I have not succeeded in observing any deviation of the X-rays even in very strong magnetic fields.

The deviation of kathode rays by the magnet is one of their peculiar characteristics; . . .

As the result of many researches, it appears that the place of most brilliant phosphorescence of the walls of the discharge-tube is the chief seat whence the X-rays originate and spread in all directions; that is, the X-rays proceed from the front where the kathode rays strike the glass. If one deviates the kathode rays within the tube by means of a magnet, it is seen that the X-rays proceed from a new point, *i.e.* again from the end of the kathode rays.
again from the end of the kathode rays.

Also for this reason the X-rays, which are not deflected by a magnet, cannot be regarded as kathode rays which have passed through the glass, for that passage cannot, according to Lenard, be the cause of the different deflection of the rays. Hence I conclude that the X-rays are not identical with the kathode rays, but are produced from the kathode rays at the glass surface of the tube.

The rays are generated not only in glass. I have obtained them in an apparatus closed by an aluminium plate 2 mm. thick. I purpose later to investigate the behaviour of other substances.

The justification of the term "rays," applied to the phenomena, lies partly in the regular shadow pictures produced by the interposition of a more or less permeable body between the source and a photographic plate or fluorescent screen.

I have observed and photographed many such shadow pictures. Thus, I have an outline of part of a door covered with lead paint; the image was produced by placing the discharge-tube on one side of the door, and the sensitive plate on the other. I have also a shadow of the bones of the hand (Figure 6), of a wire wound upon a bobbin, of a set of weights in a box, of a compass card and needle completely enclosed in a metal case. . . .

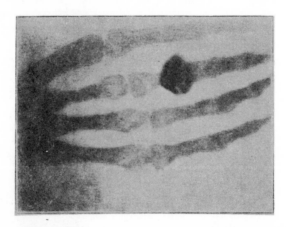

FIG. 6—Photograph of the bones in the fingers of a living human hand. The third finger has a ring upon it.

A kind of relationship between the new rays and light rays appears to exist: at least the formation of shadows, fluorescence, and the production of chemical action point in this direction.

Soon after Roentgen's paper had become widely circulated, the French scientist Becquerel became intrigued with the observation that the X-rays seemed to originate from the fluorescent spot where the cathode rays struck the end of the tube. Perhaps other types of fluorescence or phosphorescence can produce X-rays (fluorescence is the emission of light by a substance while being bomarded by an exterior stimulus, while phosphorescence is the emission of light after the external stimulus is removed). Becquerel began experiments which involved exposing fluorescent and phosphorescent minerals to the sun, and then placing them on covered photographic plates to see if X-rays would be produced which would expose the film. It turns out that fluorescent and phosphorescent minerals don't produce X-rays. Bequerel, however, thought that he demonstrated that X-rays were produced when he placed a uranium-bearing mineral on a photographic plate and found that the plate soon became exposed. In a series of other experiments, he determined that the plates were actually exposed by some new radiations which were emitted from uranium, and were not connected with fluorescence at all. This new phenomena was soon named radioactivity.

In the first report (3), Becquerel wrongly attributed the exposure of the film to phosphorescence.

With the potassium-uranium sulfates, which I have some crystals of in the form of a thin transparent crust, I was able to do the following experiment:

One wraps a photographic plate of bromide gelatin in two thick sheets of black paper so that the plate does not become clouded on being exposed to sunlight, for one day. Then place outside the paper a plate of phosphorescent substance, to expose to sunlight for several hours. When the photographic plate is subsequently developed, one sees the black silhouette of the phosphorescent substance on the negative. If one places a coin, or metallic screen pierced in a design, between the phosphorescent substance and the paper, the image of these objects appears on the negative.

One can repeat these experiments including a glass plate interposed between the phosphorescent substance and the paper, thus excluding the possibility that solar rays heat the substance causing it to vaporize, so that the effects are due to chemical action of the vapor.

One must conclude from these experiments that the phosphorescent substance in question emits radiations which penetrate opaque paper and reduce the silver salts.

In the next paper Becquerel accidentally discovered that it is not necessary to place the uranium mineral in sunlight in order for it to expose the plates. He still tended to attribute the exposure to some kind of phosphorescence, but the last paragraph presented indicates that he was not at all sure (4).

At the last meeting I presented a summation of the experiments that I have conducted to establish evidence for the invisible radiations emitted by certain phosphorescent substances, radiations which pass through objects opaque to light. . . .

The experiments I am reporting have been done with radiations emitted by crystallized lamellae of the double sulfate of uranium and potassium. $K(UO)SO_4 + H_2O$ . . .

One may very simply verify that the radiations emitted by this substance when it is exposed to sun or diffuse daylight pass through not only sheets of black paper, but also through different metals, for example, a plate of aluminum and a thin sheet of copper. . . .

I particularly insist on the following fact, that is to me a most important fact, and not what one would expect to observe: these same crystals placed facing the photographic plates, under the same conditions, and through the same screens, but screened from the excitation of incident radiation and kept in the dark, produce the same photographic impressions. Here is how I happen to have made this observation: among the preceding experiments some were prepared on Wednesday the 26th and Thursday the 27th of February, and since on those days the sun shone only intermittently, I saved the experiments I had prepared and returned them to darkness in a drawer while leaving in place the crystals of the uranium salt. Since the sun did not shine on the following days, I developed the plates on the first of March expecting to find very feeble images. To the contrary, the silhouettes appeared with great intensity. I thought that the action had continued in the dark and so proposed the following experiment.

At the bottom of an opaque box I placed a photographic plate. Then on the sensitive side I placed a lamella of the uranium salt, a convex lamella which did not touch the bromide gelatin except at a few points. Then to the side on the same plate, I arranged another lamella of the same salt, separated from the bromide gelatin by a thin sliver of glass. This operation was carried out in a dark room where the box was sealed, placed in another box, and then placed in a drawer.

I did the same thing with a frame enclosed by a plate of aluminum in which I put a photographic plate, then on the outside a lamella of the uranium salt. All of this was enclosed in an opaque box and placed in a drawer. After five hours, I developed the plates and silhouettes of the lamellae appeared in black as in the preceding experiments as if they had been rendered phosphorescent by light. . . .

An hypothesis which naturally presents itself is to suppose that radiations, whose effects have great similarity to those studied by Lenard and Roentgen, are invisible radiations of a duration infinitely longer than that of the luminous radiations emitted by the same bodies. All the same, the present experiments, while not contradicting this hypothesis, do not prescribe it either . . .

In a third paper (5), Becquerel still defended the idea that phosphorescence produced the exposure of the photographic plates, but then presented an experiment which essentially refutes this!

I will report on another interesting experiment. It is known that nitrate of uranium ceases to be phosphorescent or fluorescent when it is dissolved or melted in its water of crystallization. I then took a crystal of this salt, and after putting it in a small tube, closed by a thin plate of glass, heated it in darkness, having managed to avoid even the radiations from the alcohol lamp which I had used to heat it. When the salt had melted I allowed it to crystallize. Then I placed it on a photographic plate covered with black paper, all the while keeping the salt from exposure to light. One would not expect to observe any exposure, since all luminous excitation had been prohibited from the time that the salt ceased to be phosphorescent. Nevertheless, the impressions were as strong as for salts exposed to light, and where the salt was in contact with the plate, the impression was stronger than that of a sample of uranium sulfate placed on the same plate for comparison. . . .

Anomalies Presented by Various Substances

The uranium salts emit invisible radiations with a remarkable consistency, but this is not the case with other phosphorescent substances.

Finally (6) Becquerel became convinced that the element uranium provides the radioactivity, and not the minerals or any mineral properties of normal fluorescence or phosphorescence.

A few months ago I showed that uranium salts emit radiations whose existence has never before been recognized, and that these emissions have remarkable properties which are somewhat comparable to the properties of radiations studied by Mr. Roentgen. The radiations of the uranium salts are emitted not only when the substances are exposed to light, but also when they are kept in the dark. For two months fragments of different salts, kept isolated from the excitation of all known radiation, continue to emit new radiations, almost without appreciable weakening.* From the third of March to the third of May, these substances have been shut up in an opaque cardboard box. Since May 3rd they have been in a double box of lead which has not left the darkened room. A simple arrangement allows a photographic plate to be slipped under a black paper stretched parallel to the bottom of the box on which are placed the subjects of the experiment. The substances are not exposed to any radiation which does not penetrate the lead. Under these conditions the substances studied continue to emit active radiations. . . .

All the uranium salts I have studied, which were phosphorescent or not on exposure to light, crystallized, melted or dissolved, gave comparable results. This leads me to think that the effect was due to the presence of uranium in the salts, and that the metal should give more intense effects than its compounds.

The experiment conducted a few weeks ago with a commercial uranium powder which has been in my laboratory for a long time has confirmed this prediction: that the photographic effect is notably stronger than the impression produced by one of the uranium salts, in particular, the potassium uranium sulfate. . . .

Although I am continuing to study these new phenomena, I think that it is not without interest to point out that the emission produced by uranium is, I think, the first example of a metal presenting a phenomenon of the type of an invisible phosphorescence.

The study of these rays from uranium was taken up by Madame Curie, and later by her husband. The Curies measured the strength of the radioactivity (as Madame Curie was to name these emanations), or Becquerel rays, by recording the ionization of the air resulting from the radiations. In the first paper quoted (14) Madame Curie demonstrated that thorium, as well as uranium, is radioactive, and suggested the existence of other radioactive elements.

I have studied the conductivity of air under the influence of rays from uranium, discovered by Becquerel, and I have inquired whether substances other than those composed of uranium were capable of rendering air a conductor of electricity. . . .

All of the uranium compounds studied are active and are, in general more active if they contain more uranium.

---

*(actually, there is no weakening at all)

The thorium compounds are very active. Thorium oxide is even more active than metallic uranium.

It is noteworthy that the two most active elements, uranium and thorium, are those which have the greatest atomic weight. . . .

The minerals which have proved active all contain active elements. Two uranium minerals—pitchblende (uranium oxide) and chalcolite (copper uranyl phosphate)—are far more active than uranium itself. This fact is very remarkable and indicates that these minerals may contain an element which is much more active than uranium.

The Curies proceeded to search for new radioactive elements, and subsequently, announced the discovery of one (15). The element they had discovered was chemically very similar to bismuth, and was hundreds of times as radioactive as uranium.

Certain minerals containing uranium and thorium (pitchblende, chalcolite, uranite) are very active with regard to the emission of Becquerel rays. In an earlier work, one of us has shown that their activity is even greater than that of uranium and of thorium, and has voiced the opinion that this effect was due to some other very active substance present in small amounts in these minerals. . . .

It thus becomes very likely that if certain minerals are more active than uranium and thorium, it is because they contain a substance more active than these metals. We have tried to isolate this substance in pitchblende, and experience has confirmed the preceding predictions. . . .

The pitchblende which we have analyzed was approximately two-and-a-half times more active than the uranium in our parallel-plate apparatus. We have attacked it with acids, and we have treated the solution obtained with hydrogen sulfide. Uranium and thorium remained in the solution. We have established the following facts:

The precipitated sulfides contain a very active substance together with lead, bismuth, copper, arsenic and antimony.

This substance is completely soluble in ammonium sulfide, which distinguishes it from arsenic and antimony.

The substances which are insoluble in ammonium sulfide are soluble in nitric acid, the active substance perhaps, could be incompletely separated from lead by sulphuric acid. By separating the lead sulfate with sulphuric acid, one can in great part, dissolve the active substance brought along with the lead sulfate.

The active substance now found in solution with the bismuth and the copper is completely precipitated by ammonia, which separates it from copper.

Finally, the active material remains with the bismuth.

We have not yet discovered any exact procedure for separating the active substances from bismuth by wet methods. However, we have effected some incomplete separations based upon the following facts:

In the solution of sulfides by nitric acid, the most easily dissolved parts are the least active.

In the precipitation of the salts by water, the first parts precipitated are by far the most active.

We have observed that in heating pitchblende one obtains by sublimation very active products. This observation led us to a process of separation based on the difference in volatility between active sulfide and the bismuth sulfide. One heats the sulfides in a vacuum in a tube made of Bohemian glass at about 700°. The active sulfide deposits itself in the form of a black glazing in the parts of the tube which are at 250°–300°, while the bismuth sulfide remains in the hotter parts.

In effecting these diverse operations, one obtains products of greater and greater activity. Finally, we have obtained a substance the activity of which is about 400 times greater than that of uranium. . . .

We believe then that the substance which we have extracted from the pitchblende contains a metal not recognized before, similar to bismuth in its analytical properties. If the existence of this new metal is confirmed, we propose to call it *polonium*, after the name of the homeland of one of us.

A few months later they announced the discovery of yet another radioactive element (16).

Two of us have demonstrated that, by purely chemical procedures, one can extract from pitchblende a strongly radioactive substance. This substance is related to bismuth by its analytical properties. We have expressed the opinion that pitchblende perhaps contained a new element, for which we have proposed the name *polonium.*

The research which we are presently pursuing accords with the first results obtained; but, in the course of these investigations, we have encountered a second substance, strongly radioactive and distinguished from the first by its chemical properties. In effect, the polonium is precipitated in an acid solution by hydrogen sulfide; these salts are soluble in acids, and water precipitates them from these solutions; the polonium is completely precipitated by ammonia.

The new radioactive substance that we have just found has similar chemical properties to pure barium. . . .

Finally, this substance gives the easily recognized spectrum of barium.

We nevertheless believe that this substance, even though mainly constituted of barium, contains as well a new element which gives it radioactivity and which, furthermore, is very closely related to barium by its chemical properties. . . .

We have thus obtained chlorides having an activity 900 times greater than that of uranium. . . .

M. Demarcay has offered to examine the spectrum of our substance, with a kindness which we cannot thank too much. The results of his examination are expressed in a special note following ours. M. Demarcay has found in the spectrum a line which seems due to no known element. This line, scarcely visible with the chloride 60 times more active than uranium, has become notable with the chloride enriched by fractionation up to an activity 900 times that of uranium. The intensity of this line grows, then, at the same time as the radioactivity. This, we think, is a very serious reason for attributing it to the radioactive part of our substance.

The diverse reasons which we have listed lead us to believe that the new radioactive substance contains a new element, to which we propose to give the name *radium*. . . .

The new radioactive substance contains certainly a very large proportion of barium; nevertheless, the radioactivity is considerable. The radioactivity of radium must then be enormous.

Accompanying the preceding paper was the following note by Demarcay (17). For years it had been known that each element, when heated to incandescence in a gaseous state, emits specific colors or frequencies of electromagnetic radiation, and that this spectrum of radiation could be used to identify the element. Since the spectrum for each element is unique, it acts as "fingerprints," and was, perhaps, the most reliable way then known to identify the presence of a specific element.

M. and Mme. Curie have asked me to examine, with regard to the spectrum, a substance containing mostly barium chloride in which they suppose, for reasons developed elsewhere, the presence of a new element. This substance, dissolved in distilled water weakly acidified with HCl and submitted to the action of a spark from my coil of heavy wire, furnished a brilliant spectrum which was photographed. I have thus prepared two negatives with two exposure times, one twice the other. These two negatives, gave, by the close intensity of their lines, identically the same result. I have measured them and have been able to see: . . .

A *prominent line,* stronger than the weak lines of barium, having for λ : 3814.8 (Rowland's scale). This line does not appear to me to be attributable to any known element; first, because one cannot distinguish on the negatives in question any other lines than those already enumerated, except for some weak lines of air, which excludes all other elements which have, at the most, weak lines in the neighborhood of 3814; second, furthermore, since the mode of purification used for the substance excludes precisely the presence of those which might cause it (Fe, Cr, Co, Ni . . .) and which, besides, do not manifest themselves by any line, weak or strong. Neither the barium nor the lead, gives, as I have assured myself, any lines at all which coincide with it.

It was measured in comparison to two lines of platinum 3818.9 and 3801.5 which fall on either side of it. It is near to and distinct from a moderately intense line of bismuth.

CONCLUSION: The presence of the line 3414.8 confirms the existence, in a small quantity of a new element in the barium chloride of M. and Mme. Curie.

The remainder of this chapter deals with the isolation and identification of the radiations emitted by radioactive elements, and the experimental procedures which led to an explanation of radioactivity.

By 1899, workers in Germany, France and England were attempting to discover the nature of radioactive emanations. Early that year Rutherford demonstrated that these radiations are not homogenous, but consist of at least two different types. He began his publication of this discovery with a brief summation of the knowledge of these rays accumulated up to that date (31).

The remarkable radiation emitted by uranium and its compounds has been studied by its discoverer, Becquerel, and the results of his investigations on the nature and properties of the radiation have been given in a series of papers in the *Comptes Rendus*. He showed that the radiation, continuously emitted from uranium compounds, has the power of passing through considerable thicknesses of metals and other opaque substances; it has the power of acting on a photographic plate and of discharging positive and negative electrification to an equal degree. The gas through which the radiation passes is made a temporary conductor of electricity and preserves its power of discharging electrification for a short time after the source of radiation has been removed. . . .

Complex Nature of Uranium Radiation.

Before entering on the general phenomena of the conduction produced by uranium radiation, an account will be given of some experiments to decide whether the same radiation is emitted by uranium and its compounds and whether the radiation is homogeneous. . . .

Rutherford employed an experimental procedure which measured the amount of radiation absorbed by successively thicker layers of aluminum.

It will be observed* that for the first three layers of aluminium foil, the intensity of the radiation falls off according to the ordinary absorption law, and that, after the fourth thickness, the intensity of the radiation is only slightly diminished by adding another eight layers.

Number of Layers of Aluminum Foil.	Leak per Minute in Scale-divisions.	Ratio.
0	182	
1	77	•42
2	33	•43
3	14•6	•44
4	9•4	•65
12	7	

The aluminium foil in this case was about •0005 cm. thick, so that after the passage of the radiation through •002 cm. of aluminum the intensity of the radiation is reduced to about 1/20 of its value. The addition of a thickness of •001 cm. of aluminum has only a small effect in cutting down the rate of leak. The intensity is, however, again reduced to about half of its value after passing through an additional thickness of •05 cm., which corresponds to 100 sheets of aluminium foil.

These experiments show that the uranium radiation is complex, and that there are present at least two distinct types of radiation—one that is very readily absorbed, which will be termed for convenience the α radiation, and the other of a more penetrative character, which will be termed the β radiation. . . .

As far, however, as the experiments have gone, the results seem to point to the conclusion that the β radiation is approximately homogeneous, although it is possible that other types of radiation of either small intensity or very great penetrating power may be present. . . .

*(see table)

> The cause and origin of the radiation continuously emitted by
> uranium and its salts still remain a mystery. All the results that have
> been obtained point to the conclusion that uranium gives out types
> of radiation which, as regards their effect on gases, are similar to
> Röntgen rays.

That same year, Becquerel and others, found that the β rays of
Rutherford could be deviated in a magnetic field in the same direction as cathode ray particles, and hence, were negatively charged
particles. Within a year e/m ratios for β particles strongly suggested
that β particles were in fact identical to cathode ray particles (electrons). Discovering the identity of the α particle was a more difficult
and protracted problem which we will follow very closely. A third
type of radiation from radioactive elements termed γ rays, which did
not deviate in a magnetic field, was discovered in 1900.

Before we follow the identification of the radioactive emissions
further, however, we should first look at the explanation of radioactivity which was set forth by Rutherford and Soddy in 1902 and
1903. In 1900, Sir William Crookes had discovered that he could
remove from uranium, by chemical means, a substance which was
different from uranium and seemed to contain most of the radioactivity previously attributed to uranium. In other words, after the
uranium X, as he called the new substance, was removed from the
uranium, the uranium ceased temporarily to be radioactive. In 1901
and 1902, Rutherford along with Owens and Soddy, had been studying the radioactivity of thorium. They found that thorium produced
a radioactive gas which they called "thorium emanation". In papers
published earlier than those to be presented here, Rutherford and
Soddy announced the discovery of a radioactive "thorium X," chemically separable from thorium, which actually seemed responsible for
the production of the thorium emanation. This led them to conclude
that thorium was actually changing into a new substance. This suggests a transmutation of one element into another. The alchemists'
dream of transmutation appeared to be a naturally occurring process
among radioactive elements!

In the first paper, excerpts of which appear below, they announced these results (50).

> The following papers give the results of a detailed investigation of
> the radioactivity of thorium compounds which has thrown light on
> the questions connected with the source and maintenance of the

energy dissipated by radioactive substances. Radioactivity is shown to be accompanied by chemical changes in which new types of matter are being continuously produced. These reaction products are at first radioactive, the activity diminishing regularly from the moment of formation. Their continuous production maintains the radioactivity of the matter producing them at a definite equilibrium-value. The conclusion is drawn that these chemical changes must be subatomic in character.

The present researches had as their starting point the facts that had come to light with regard to thorium radioactivity . . . .

Besides being radioactive in the same sense as the uranium compounds, the compounds of thorium continuously emit into the surrounding atmosphere a gas which possesses the property of temporary radioactivity. This "emanation," as it has been named, is the source of rays, which ionize gases and darken the photographic film. . . .

They then summarized their experimental results.

Summary of Results.

The foregoing experimental results may be briefly summarized. The major part of the radioactivity of thorium—ordinarily about 54 per cent—is due to a non-thorium type of matter, ThX, possessing distinct chemical properties, which is temporarily radioactive, its activity falling to half value in about four days. The constant radioactivity of thorium is maintained by the production of this material at a constant rate. Both the rate of production of the new material and the rate of decay of its activity appear to be independent of the physical and chemical condition of the system. . . .

Now came the bombshell: transmutation of one element into another!

General Theoretical Considerations.

Turning from the experimental results to their theoretical interpretation, it is necessary to first consider the generally accepted view of the nature of radioactivity. It is well established that this property is the function of the atom and not of the molecule. Uranium and thorium, to take the most definite cases, possess the property in whatever molecular condition they occur, and the former also in the elementary state. So far as the radioactivity of different compounds of different density and states of division can be compared together, the intensity of the radiation appears to depend only on the quantity of active element present. . . .

All the most prominent workers in this subject are agreed in considering radioactivity an atomic phenomenon. M. and Mme. Curie, the pioneers in the chemistry of the subject, have recently put forward their views (*Comptes Rendus,* cxxxiv. 1902, p. 85). They state that this idea underlies their whole work from the beginning and created their methods of research. M. Becquerel, the original discoverer of the property for uranium, in his announcement of the recovery of the activity of the same element after the active constituent had been removed by chemical treatment, points out the significance of the fact that uranium is giving out cathode rays. These, according to the hypothesis of Sir William Crookes and Prof. J.J. Thomson, are *material* particles of mass one thousandth of the hydrogen atom.

Since, therefore, radioactivity is at once an atomic phenomenon and accompanied by chemical changes in which new types of matter are produced, these changes must be occurring within the atom, and the radioactive elements must be undergoing spontaneous transformation. The results that have so far been obtained, which indicate that the velocity of this reaction is unaffected by the conditions, makes it clear that the changes in question are different in character from any that have been before dealt with in chemistry. It is apparent that we are dealing with phenomena outside the sphere of known atomic forces. Radioactivity may therefore be considered as a manifestation of subatomic chemical change.

In their next paper (51) Rutherford and Soddy sharpened their interpretation of the experimental results and added the idea that radioactive emanations (α and β particles, and γ rays) are emitted *during* the transformation or transmutation of an atom of one element into an atom of another.

Further Theoretical Considerations.

Enough has been brought forward to make it clear that in the radioactivity of thorium, and, by analogy, of radium, we are witnessing the effect of a most complex series of changes, each of which is accompanied by the continous production of a special kind of active matter. The complexity of the phenomenon gives rise to an important question concerning the fundamental relation between the changes which occur and radioactivity. So far it has been assumed, as the simplest explanation, that the radioactivity is *preceded* by chemical change, the products of the latter possessing a certain amount of available energy dissipated in the course of time. A slightly different view is at least open to consideration, and is in some ways preferable. Radioactivity may be an *accompaniment* of the change, the amount of the former at any instant being proportional to the amount of the latter. On this view the nonseparable radioactivities of thorium and uranium would be caused by the primary change in which ThX and UrX are produced. The activity of ThX would be caused by the secondary change producing the emanation, the activity of the emanation by a tertiary change in which the matter causing the excited activity is produced, the activity of the latter being derived from still further changes.

In still another joint publication the following year, Rutherford and Soddy discussed the energy involved in radioactive transformation and made prophetic suggestions as to its importance (52).

It has been pointed out that these estimates* are concerned with the energy of radiation, and not with the total energy of radioactive change. The latter, in turn, can only be a portion of the internal energy of the atom, for the internal energy of the resulting products remains unknown. All these considerations point to the conclusion that the energy latent in the atom must be enormous compared with that rendered free in ordinary chemical change. Now the radio-elements differ in no way from the other elements in their chemical and physical behaviour. On the one hand they resemble chemically their inactive prototypes in the periodic system very closely, and on the other they possess no common chemical characteristic which could be associated with their radioactivity. Hence there is no reason to assume that this enormous store of energy is possessed by the radio-elements alone. It seems probable that atomic energy in general is of a similar, high order of magnitude, although the absence of change prevents its existence being manifested. The existence of this energy accounts for the stability of the chemical elements as well as for the conservation of radioactivity under the influence of the most varied conditions. It must be taken into account in cosmical physics. The maintenance of solar energy, for example, no longer presents any fundamental difficulty if the internal energy of the component elements is considered to be available, *i.e.*, if processes of subatomic change are going on.

These rather radical explanations of radioactivity were not immediately accepted by many physicists. This is well illustrated in a letter by Rutherford (34) in which he argues against alternative explanations by Mme. Curie on thorium "emanation" and related radioactive phenomena.

... I have recently shown that it is extremely probable that the greater proportion of the radiation from the emanation is mate-

*(of the energy of radioactive emissions)

rial in nature, and consists of heavy charged bodies projected with
great velocity, whose mass is of the same order as that of the hydrogen atom. In view of these results, which so strongly confirm the
theory of the material nature of the emanation, the alternative
theory proposed by M.P. Curie that the emanation consists of
"centres de condensation d'énergie situés entre les molécules du gaz et
qui peuvent être entraînés avec lui," appears to me unnecessary.

The interesting result, obtained by M. Curie of the exponential
law of decay of the radium emanation under all conditions, is only
one of many others that have now been accumulated. I quite agree
with M. Curie that such results cannot be satisfactorily explained on
the laws of *ordinary* chemical change, but the difficulty disappears
on the view already put forward by Mr. Soddy and myself . . .
that the radioactivity of the elements is a manifestation
of *subatomic chemical change*, and that the radiations accompany
the change.

There is no *à priori* reason to suppose that temperature would
affect the rate of atomic disintegration; in fact the general experience of chemistry in failing to transform the elements is distinctly
opposed to such a view. It is therefore not surprising that, if radioactivity is an accompaniment of subatomic change, the process
should be independent of the ordinary chemical and physical agents
at our disposal.

The following excerpts, from a paper by Soddy (54) in 1906, again
illustrated the controversial aspects of the new atomic doctrine. Lord
Kelvin led the attack against the Rutherford-Soddy transmutation
theory of radioactivity. By this time, it had been demonstrated that
helium was one of the products of the radioactive decay of radium,
and it is on this point that the argument opens. The excerpts also
show the more human side of the scientists with the quarrels and
rivalries between the chemists and the physicists. Lastly these paragraphs indicate fairly well the experimental basis and the degree of
acceptance of the transmutation theory of radioactivity by 1906. It
is a very interesting and informative paper.

Lord Kelvin's opening challenge (August 9) was broad and
sweeping. He took exception to the statement, made by the writer in
opening the discussion on the evolution of the elements at the
British Association at York, that the production of helium from
radium has established the fact of the gradual evolution of one element into others, and denied that this discovery affected the atomic
doctrine . . .

The challenge was taken up on the other side successively by
Sir Oliver Lodge, the Hon. Mr. Strutt, and other well-known authori-

ties, and it soon became apparent that for argument at least Lord Kelvin on his side had to rely practically on himself alone. Prof. Armstrong, it is true, immediately enrolled under Lord Kelvin's banner, and entered the lists with an embracing criticism of physicists in general whom, he declared, are strangely innocent workers under the all-potent influence of formula and fashion. He made the statement that no one had handled radium in such quantity or in such manner that we can say precisely what it is, and throughout put the word *radium* in inverted commas. . . .

Prof. Armstrong's letter merely served to provide Sir Oliver Lodge with justification for his favourite theme, which appears to be that whereas chemists have an instinct of their own for arriving at their results, reason is the monopoly of the physicist, whose results the chemist usually manages to absorb in the end. No better argument against the unfairness of this could be provided than by the history of radioactivity itself, which owes at least as much to the chemist as to the physicist. Prof. Armstrong is almost alone among chemists, as Lord Kelvin is among physicists, in his hostility to the new doctrines. . . .

Lord Kelvin in his replies (August 20 and 24) made it clear that he accepted as a fact the continuous evolution of helium from radium, and this admission narrowed very much the issue involved. . . .

For helium is produced from the emanation of radium, about which no question of its being really reproduced can exist. For the removal of the emanation is marked by changes in the radioactivity, notably by the β rays, which vanish when the emanation is removed. The recovery of the radioactivity occurs at a definite rate, and is concomitant to the reproduction of emanation, which can at any time be again extracted as before. As there is no question of the radium creating helium, the only point open for argument is the exact character of the decomposition by which it and the emanation which gives rise to it are formed. As there was no further reply to this criticism, it may be taken that the main point of the disintegration theory that there is a continuous change in the radioactive matter accompanying the radioactivity, is unanswerable. . . .

He* quoted a statement of Prof. Rutherford in favour of regarding radium as chemical compound of helium and other elements, and suggested that radium might be made up of one atom of (?) lead and four of helium. In a final letter (September 4) Sir Oliver Lodge pointed out that this was the key of the position. Is radium a compound or an element? It is satisfactory that, after so much fencing with the question, so simple an alternative has been arrived at. Perhaps the most significant thing about the view that radium is a

*(Kelvin)

compound is the silence of the chemists. Surely a chemist might fairly be supposed to know whether a change is what is called a chemical change or not, and the fact that it has been left to a physicist to adopt this view seems fair comment. Not even Prof. Armstrong has yet accepted it. . . .

The theory that radium is a compound, waiving the qualification *chemical*, will no doubt serve sufficiently well for the present as a point of common agreement. . . .

It expresses a bare minimum of established fact which even the most sceptical are unable to invalidate. This minimum, briefly stated, is that radium is undergoing a continuous change intimately connected with its radioactivity, and that in this change helium is produced, and an enormous but definite amount of energy liberated. . . .

But what a miserable fraction, even of the known facts, this minimum is! Ostensibly an explanation of radioactivity, it begins and ends with the fact of the gradual evolution of helium from radium. The numerous other products of radium, the volatile emanation and its nonvolatile products, known by their characteristic radioactivity, much as minute quantities of ordinary gases and solids are known by their characteristic spectra, the slower changing later products, of which polonium is one, and is chemically as reminiscent of tellurium as its parent is of barium, remain still to be systematically accounted for. On the important subject of the nature of the α, β, and γ rays, and their origin, the view is silent! The fact is ignored that radioactivity is, to use Mme. Curie's happy expression, an atomic property, that is, is independent of the particular state of chemical combination of the radioelement. . . .

It is the glory of the accepted view that it harmonises and correlates all the preceding problems, offering a simple and unstrained physical explanation of each, capable of being tested by quantitative experiment. In addition, it reaches out in every direction in broad, bold predictions, a few of which, like the production of helium from radium and the constancy of ratio between uranium and radium in minerals, have been brilliantly confirmed by experiment, while the majority simply await more refined experimental methods of attack. Of what other theory could the remark be made, which is attributed to Prof. Rutherford, that when a single experimental fact is established which does not conform to the disintegration theory it will be time to abandon it? The theory would have to be fundamental indeed to pass this test. . . .

Give a scientific man a few milligrams of radium in solution and ask him to perform for himself some of the stock experiments with the emanation, for example, its condensation by liquid air, the concentration on the negative electrode of the active deposit formed by it, the steady decay of its powers after removal from the radium, and the growth of new emanation by the radium, kept, let us say, in another building or another country; then the radium emanation passes from being a phrase to a fact which no theory can safely ignore. The same is equally true of thorium X, radium C, and the numerous other successive products of radioactive change.

It would be a pity if the public were misled into supposing that those who have not worked with radioactive bodies are as entitled to as weighty an opinion as those who have. The latter are talking of facts they know, the former frequently of terms they have read of. If, as a result of the recent controversy, it has been made clear that atomic disintegration is based on experimental evidence, which even its most hostile opponents are unable to shake or explain in any other way, the best ends of science will have been served. The sooner this is understood the better, for in radioactivity we have but a foretaste of a fountain of new knowledge, destined to overflow the boundaries of science and to impregnate with teeming thought many a high and arid plateau of philosophy.

We now return to the systematic experimental steps which were taken in order to determine the identity of the rays emitted by radioactive substances. By 1900 Becquerel had demonstrated the similarities between β particles and electrons, but the identity of γ rays and the α particle were still largely speculative. We shall follow closely the work on the α particle for two reasons. First, the systematic determination of the α particle's identity is a magnificent example of an experimentally-based scientific process. Second, the α particle was to become an all important tool in further discoveries of the nature of the atom.

In 1903 Rutherford summed up the knowledge concerning the α and β particles, and γ radiation (32).

It is known that uranium, thorium, and radium emit two types of radiation. One type is not appreciably deviable by a magnetic or an electric field, and is very easily absorbed in matter. These will be called the α rays. The others are deviable and more penetrating in character, and will be called the β rays. In addition I have shown that thorium and radium emit some rays nondeviable in character, but of very great penetrating power. . . .

It has been difficult to offer a satisfactory explanation of the nature of these rays. I have previously shown as untenable the view that they are secondary rays due to the emission of β rays. I have been recently led, by a mass of indirect evidence, to the view that the α rays are in reality charged bodies projected with great velocity. The ionizing effect of the rays is due to the collision of the projected body with the molecules of the gas, in the same way that the cathode rays ionize the gas in their path. . . .

The evidence in favour of the projection nature of the α rays is so far all indirect in character, and is briefly summarized below:—

(1) The absorption of the α rays in matter (like the β rays which we know are projected particles) is approximately proportional to the density of the material. . . .

(2) The absorption of the α rays by a given thickness of matter *increases* rapidly with the thickness traversed.

I have found that this is a general property of the α radiations not only for the radioactive elements proper, but for the radiations from the emanation and excited bodies. This is to be expected if the rays consist of projected particles, but is difficult to explain if the radiations are æther-waves similar to Röntgen rays. . . .

If these rays are due to projected charged particles they should possess the properties of the α* rays of deflectability by a magnetic and electric field.

No deviation of the α rays has so far been detected in a strong magnetic field, but the experiments have not yet been pushed to the necessary limit. The results, however, indicate that if the rays are deflectable, the deviation is minute compared with the β rays. This is to be expected if the mass of the expelled particle is large compared with the electron.

Only a month later Rutherford demonstrated that α particles could be deflected in a magnetic field. These deflections indicated that α particles were positively charged (33).

Radium gives out three distinct types of radiation:—

(1) The α rays, which are very easily absorbed by thin layers of matter and which give rise to the greater portion of the ionization of the gas observed under the usual experimental conditions.

(2) The β rays, which consist of negatively charged particles projected with high velocity, and which are similar in all respects to cathode rays produced in a vacuum tube.

(3) The γ rays, which are nondeviable by a magnetic field, and which are of a very penetrating character.

These rays differ very widely in their power of penetrating matter. The following approximate numbers, which show the thickness of aluminium traversed before the intensity is reduced to one-half, illustrate this difference.

Radiation.	Thickness of Aluminium.
α rays	·0005 cm.
β rays	·05 cm.
γ rays	8 cms.

*(β)

In this paper an account will be given of some experiments which show that the α rays are deviable by a strong magnetic and electric field. The deviation is in the opposite sense to that of the cathode rays, so that the radiations must consist of positively charged bodies projected with great velocity. . . .

In his experimental procedure, illustrated below, Rutherford measured the flow of α particles by measuring the ionization produced by the particles in air. The greater the flow of α particles, the faster the leaves of the electroscope come together.

Magnetic Deviation of the Rays.

Figure 7 shows the general arrangement of the experiment. The rays from a thin layer of radium passed upwards through a number of narrow slits, G, in parallel, and then through a thin layer of aluminium foil ·00034 cm. thick into the testing vessel V. The ionization produced by the rays in the testing vessel was measured by the rate of movement of the leaves of a gold leaf electroscope B . . .

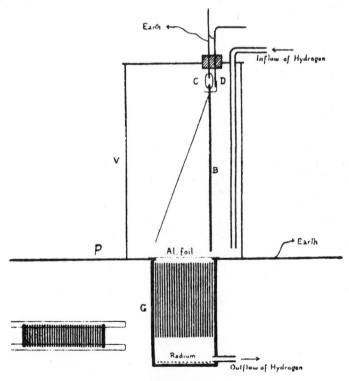

Figure 7.

When the magnetic field is turned on, any deviation of the α particles will drive them into the walls of the slits (G) so that they cannot pass into the electroscope chamber and ionize the air. In the graph presented, Rutherford showed that a deviation of the α particles does occur which is proportional to the strength of the magnetic field. Then, in a very clever procedure, he demonstrated that the α particle must be positively charged.

The curve (Figure 8) shows that the amount deviated is approximately proportional to the magnetic field.

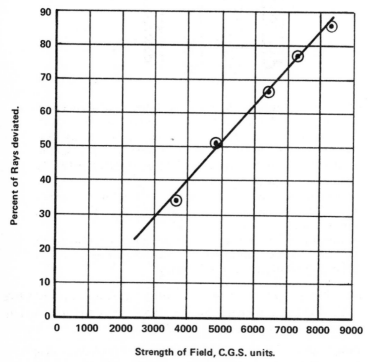

Strength of Field, C.G.S. units.

Figure 8

With another apparatus, with a mean air space of ·055 cm., the rays were *completely* deviated by a uniform magnetic field of strength 8400 units extending over the length of the plates, a distance of 4·5 cms.

Direction of the Deviation of the Rays

In order to determine the direction of the deviation, the rays were passed through slits of 1 mm. width. Each slit was about half covered by a brass plate in which air spaces were cut to correspond accurately with the system of parallel plates. Figure 9 represents an

Figure 9

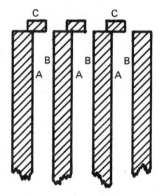

enlarged section of three of the plates, with the metal plate C half covering the slit AB. If a magnetic field is applied, not sufficiently great to deviate all the rays, the rate of discharge in the testing vessel when the rays are deviated in the direction from A to B should be much greater than when the magnetic field is reversed, *i.e.*, when the rays are deviated from B to A. This was found to be the case, for while the rate of discharge was not much diminished by the application of the field in one direction, it was reduced to about one quarter of its value by reversal of the field.

In this way it was found that the direction of deviation in a magnetic field was *opposite* in sense to the cathode rays, *i.e.*, the rays consisted of positively charged particles.

At this time other people were working on γ rays, and in 1904 Eve summed up the evidence which indicated that γ rays are not particulate, but are electromagnetic waves, similar to X-rays (18).

Direct experiments by Becquerel, Rutherford, Villard, and others have failed to detect any curvature of the γ rays in a magnetic field. There is at present no evidence that the γ rays consist of negatively charged particles. On the other hand, the β rays can be deflected completely away from the γ rays. This abrupt discontinuity in the magnetic spectrum indicates that γ rays are not merely a group of those β rays which have the highest velocity.

We may conclude that the γ rays either consist of particles practically devoid of electric charge, or are of the type of Röntgen rays, or have a special character of their own hitherto unknown.

The dissimilarity between Röntgen and γ rays in connexion with the ionization of gases and vapours, and in the character of the secondary radiations, is probably due to the fact that the Röntgen pulses are more broad than those which constitute the γ rays. The differences are decreased when harder bulbs are used, and the resulting pulses are more thin.

Theoretical reasoning still supports the view that the expulsion of the β particles must be accompanied by pulses similar to those set up on the stoppage of the cathode particles; and, until there is definite evidence to the contrary, it is reasonable to suppose that these pulses, analogous to the Röntgen rays, constitute the γ rays.

By 1905 Rutherford was attempting to identify the α particle by determining its charge to mass ratio (e/m), as Thomson had done with cathode ray particles (35).

This paper contains a preliminary account of experiments made to determine with accuracy the ratio e/m of the α particles projected from radium....

The determinations of e/m and v for different strengths of electric and magnetic fields were very concordant. The values of e/m in the different experiments lay between $5 \cdot 0 \times 10^3$ and $5 \cdot 2 \times 10^3$. A separate experiment showed that the value of e/m was unaltered by passing the rays through the mica screen....

Since the value of e/m for the hydrogen atom in the electrolysis of water is about 10^4, the present results show that, within the limit of experimental error, the apparent mass of the α particle from radium C is twice that of the hydrogen atom.

The question whether the α particle is a molecule of hydrogen, an atom of helium, or a helium molecule carrying twice the ionic charge, was discussed. It is not at present possible to decide definitely between these possibilities. Further experiments are in progress to determine the value of e/m for the α particles from the other radium products and from thorium and actinium.

Some workers at this time were already convinced that the α particle was, in fact, a helium atom minus its electrons. Soddy (53), in a discussion of the production of radium from uranium, suggests this possibility.

The continuous disintegration series thus revealed, starting with the heaviest element known, embraces the majority of the known types of radioactive matter, and although direct experimental evidence is still lacking, probably ends, so far as the manifestation of radioactive phenomena is concerned, in the ultimate production of one of the heaviest nonradioactive elements, bismuth or lead....

Thus in the change of the radium atom (225) into polonium, four α particles are expelled and a reduction of atomic weight amounting to about 13 units occurs, so that the view that the α particle is or becomes an atom of helium seems to be fairly well borne out.

However, definite experimental proof of the α particle's identity as a helium atom was several years off. Rutherford explains the nature of the evidence in 1906 (37).

We may thus reasonably conclude that the α particles expelled from the different radio-elements have the same mass in all cases. This is an important conclusion; for it shows that uranium, thorium, radium, and actinium, which behave chemically as distinct elements, have a common product of transformation. The α particle constitutes one of the fundamental units of matter of which the atoms of these elements are built up. . . .

It is now necessary to consider what deductions can be drawn from the observed value of e/m found for the α particle. The value of e/m for the hydrogen ion in the electrolysis of water is known to be very nearly 10^4. The hydrogen ion is supposed to be the hydrogen atom with a positive charge, so that the value of e/m for the hydrogen atom is 10^4. The observed value of e/m for the α particle is $5 \cdot 1 \times 10^3$, or, in round numbers, one half of that of the hydrogen atom. The density of helium has been found to be $1 \cdot 98$ times that of hydrogen, and from observations of the velocity of sound in helium, it has been deduced that helium is a monatomic gas. From this it is concluded that the helium atom has an atomic weight $3 \cdot 96$. If a helium atom carries the same charge as the hydrogen ion, the value of e/m for the helium atom should consequently be about $2 \cdot 5 \times 10^3$. If we assume that the α particle carries the same charge as the hydrogen ion, the mass of the α particle is twice that of the hydrogen atom. We are here unfortunately confronted with several possibilities between which it is difficult to make a definite decision.

The value of e/m for the α particle may be explained on the assumptions that the α particle is (1) a *molecule* of hydrogen carrying the ionic charge of hydrogen, (2) a helium atom carrying *twice* the ionic charge of hydrogen, or (3) *one half* of the helium atom carrying a single ionic charge.

The hypothesis that the α particle is a molecule of hydrogen seems for many reasons improbable. If hydrogen is a constituent of radioactive matter, it is to be expected that it would be expelled in the atomic, and not in the molecular state. In addition, it seems improbable that, even if the hydrogen were initially projected in the molecular state, it would escape decomposition into its component atoms in passing through matter, for the α particle is projected at an

enormous velocity, and the shock of the collisions of the α particle with the molecules of matter must be very intense, and tend to disrupt the bonds that hold the hydrogen atoms together. If the α particle is hydrogen, we should expect to find a large quantity of hydrogen present in the old radioactive minerals, which are sufficiently compact to prevent its escape. This does not appear to be the case, but, on the other hand, the comparatively large amount of helium present supports the view that the α particle is a helium atom. A strong argument in support of the view of a connexion between helium and the α particle rests on the observed facts that helium is produced by actinium as well as by radium. The only point of identity between these two substances lies in the expulsion of α particles of the same mass. The production of helium by both substances is at once obvious if the helium is derived from the accumulated α particles, but is difficult to explain on any other hypothesis. We are thus reduced to the view that either the α particle is a helium atom carrying twice the ionic charge of hydrogen, or is half of a helium atom carrying a single ionic charge. . . .
. . . the second * hypothesis has the merit of greater simplicity and probability

The final decisive proof was offered by Rutherford and Royds in 1909 (49). In this particularly elegant experiment they collected α particles in a small cathode ray tube in order to see if the α particles, when excited by cathode ray bombardment, would emit the same spectrum as that of helium. The spectrum produced by the trapped α particles was identical to that of helium!

The experimental evidence collected during the last few years has strongly supported the view that the α particle is a charged helium atom, but it has been found exceedingly difficult to give a decisive proof of the relation. In recent papers, Rutherford and Geiger have supplied still further evidence of the correctness of this point of view. The number of α particles from one gram of radium have been counted, and the charge carried by each determined. The values of several radioactive quantities, calculated on the assumption that the α particle is a helium atom carrying two unit charges, have been shown to be in good agreement with the experimental numbers. In particular, the good agreement between the calculated rate of production of helium by radium and the rate experimentally determined by Sir James Dewar, is strong evidence in favour of the identity of the α particle with the helium atom.

The methods of attack on this problem have been largely

*(Refers to numbered assumptions above)

indirect, involving considerations of the charge carried by the helium atom and the value of *e/m* of the α particle. The proof of the identity of the α particle with the helium atom is incomplete until it can be shown that the α particles, accumulated quite independently of the matter from which they are expelled, consist of helium. . . .

We have recently made experiments to test whether helium appears in a vessel into which the α particles have been fired, the active matter itself being enclosed in a vessel sufficiently thin to allow the α particles to escape, but impervious to the passage of helium or other radioactive products.

The experimental arrangement is clearly seen in the figure. The

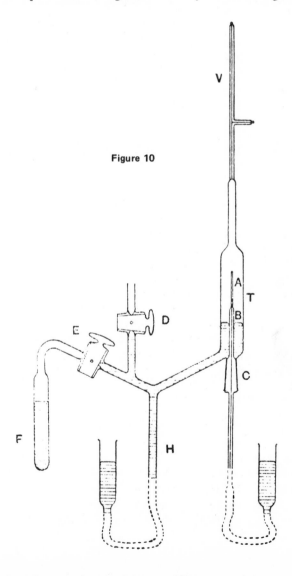

Figure 10

equilibrium quantity of emanation from about 140 milligrams of radium was purified and compressed by means of a mercury column into a fine glass tube A about 1·5 cms. long. This fine tube, which was sealed on a larger capillary tube B, was sufficiently thin to allow the α particles from the emanation and its products to escape, but sufficiently strong to withstand atmospheric pressure. . . .

The glass tube A was surrounded by a cylindrical glass tube T, 7·5 cms. long and 1·5 cms. diameter, by means of a ground glass joint C. A small vacuum tube V was attached to the upper end of T. The outer glass tube T was exhausted by a pump through the stop-cock D, and the exhaustion completed with the aid of the charcoal tube F cooled by liquid air. By means of a mercury column H attached to a reservoir, mercury was forced into the tube T until it reached the bottom of the tube A.

Part of the α particles which escaped through the walls of the fine tube were stopped by the outer glass tube and part by the mercury surface. If the α particle is a helium atom, helium should gradually diffuse from the glass and mercury into the exhausted space, and its presence could then be detected spectroscopically by raising the mercury and compressing the gases into the vacuum tube. . . .

At intervals after the introduction of the emanation the mercury was raised, and the gases in the outer tube spectroscopically examined. After 24 hours no trace of the helium yellow line was seen; after 2 days the helium yellow was faintly visible: after 4 days the helium yellow and green lines were bright; and after 6 days all the stronger lines of the helium spectrum were observed. The absence of the neon spectrum shows that the helium present was not due to a leakage of air into the apparatus. . . .

These experiments thus show conclusively that the helium could not have diffused through the glass walls, but must have been derived from the α particles which were fired through them. In other words, the experiments give a decisive proof that the α particle after losing its charge is an atom of helium.

We conclude this chapter with a brief summary of radioactivity as written by Bumstead in 1906 (10).

The study of the phenomena of radioactivity during the past five or six years, and, in particular, the brilliant series of experiments and deductions which we owe to Rutherford, have left little room for doubt that a certain proportion of the atoms of radioactive elements are continually breaking up, and that the constant emission of energy by these bodies is a result of this atomic disintegration. This process in any given radioactive body appears to be going on at a fixed and definite rate which is characteristic of the particular substance studied and which is quite uninfluenced by any external circumstances whatever. Although radioactive substances have been subjected to the greatest extremes of temperature available in the laboratory, and to great variations in other physical and chemical conditions, no certain results have been obtained (so far as the writer is aware) which point to any corresponding change in the rate of decay of the substance. In fact the process of atomic disintegration has appeared to be quite beyond human control.

Chapter 4

The Nuclear Atom and Energy Considerations

In this chapter we will follow the early development of the nuclear model of the atom which is essentially the model of the atom accepted today. The development of this model can be separated into two distinct, though interconnected, sequences of investigation.

The first of these sequences involved interpretations drawn from the physical interaction of one matter system with another. It consisted of a series of experiments, usually lumped together under the title of "gold foil experiment," which was conducted by Rutherford and his colleagues at their Manchester lab, during the first dozen years of the 20th century. The experiments stem directly from Rutherford's work with radioactivity and involved bombarding thin foils of various metals with α particles. Rutherford had previously demonstrated in his work on radioactivity, that the α particles, which each have a mass four times as great as the hydrogen atom, were released from atoms undergoing radioactive disintegration, at velocities up to 20,000 miles per second. This relatively large mass and high velocity make the α particle a rather impressive atomic "bullet," and Rutherford thought that by bombarding atoms of other elements with these "bullets," some knowledge of the internal structure of the atom might be discovered. As we shall see, he was quite correct in his thinking.

The second sequence of investigations involved the interaction of matter and energy systems. Specifically, these investigations produced an explanation of observations concerning the absorption and radiation of energy by atoms. Some of the investigations involved the release of X-rays by atoms when bombarded by cathode rays. Others involved the absorption and reradiation of other frequencies of electromatic radiation, particularly in the light range of the electromagnetic spectrum. These studies complemented Rutherford's work, and were undertaken in an attempt to more fully substantiate or explain Rutherford's nuclear model of the atom.

The 1903 paper by the Japanese physicist Nagaoka was prophetic (29). He anticipated both the nuclear model of the atom and the important role that the energy emitted by atoms was to play in establishing this model. His model is described as "Saturnian" since it is analogous to the planet Saturn surrounded by its rings. Electrostatic forces of attraction in the atom would be analogous to the gravitational forces holding together Saturn and its rings.

> Since the discovery of the regularity of spectral lines, the kinetics of a material system giving rise to spectral vibrations has been a favourite subject of discussion among physicists. . . .
>
> I propose to discuss a system whose small oscillations accord qualitatively with the regularity observed in the spectra of different elements and by which the influence of the magnetic field on band- and line-spectra is easily explicable. The system here considered is quasistable, and will at the same time serve to illustrate a dynamical analogy of radioactivity, showing that the singular property is markedly inherent in elements with high atomic weights. . . .
>
> The system, which I am going to discuss, consists of a large number of particles of equal mass arranged in a circle at equal angular intervals and repelling each other with forces inversely proportional to the square of distance; at the centre of the circle, place a particle of large mass attracting the other particles according to the same law of force. If these repelling particles be revolving with nearly the same velocity about the attracting centre, the system will generally remain stable, for small disturbances, provided the attracting force be sufficiently great. The system differs from the Saturnian system . . . in having repelling particles instead of attracting satellites. The present case will evidently be *approximately* realized if we replace these satellites by negative electrons and the attracting centre by a positively charged particle. The investigations on cathode rays and radioactivity have shown that such a system is conceivable as an ideal atom. . . .

Nagaoka then anticipated one of the most serious problems faced by any model of the atom which includes electrons moving about within the atom: electrons moving in an orbit are being constantly accelerated (change of direction) and therefore, should radiate away energy, slow down and eventually stop.

> The objection to such a system of electrons is that the system must ultimately come to rest, in consequence of the exhaustion of energy by radiation, if the loss be not properly compensated. . . .

He then proceeded with a highly mathematical determination of possible electron ring configurations. He ended the paper with the following paragraph:

> There are various problems which will possibly be capable of being attacked on the hypothesis of a Saturnian system, such as chemical affinity and valency, electrolysis and many other subjects connected with atoms and molecules. The rough calculation and rather unpolished exposition of various phenomena above sketched may serve as a hint to a more complete solution of atomic structure.

We will now examine the series of experiments which led Rutherford in 1911 to propose, as an explanation of experimental results, a model of the atom not greatly different from that of Nagaoka.

By 1906 Rutherford and his colleagues had begun bombarding substances with α particles. In the following excerpts (36) he reports his first thoughts about the scattering of α particles as they pass through solid matter.

> Since the atom is the seat of intense electrical forces, the β particle in passing through matter should be much more easily deflected from its path than the massive α particle. We know experimentally that this is the case. . . .
>
> . . . , the α particle, on account of its enormous energy of motion, plunges through the atoms of matter without suffering much deflexion from its path. As I pointed out in a previous paper, there is, however, an undoubted slight scattering or deflexion of the path of the α particle in passing through matter. . . .
>
> From measurements of the width of the band due to the scattered α rays, it is easy to show that some of the α rays in passing through the mica have been deflected from their course through an angle of about 2°. It is possible that some were deflected through a considerably greater angle; but, if so, their photographic action was too weak to detect on the plate.

Under the direction of Rutherford, Geiger and Marsden continued the scattering experiments, and investigated particularly, the possibility of deflecting α particles through angles *larger* than 2 degrees. Their results, presented in the next paper (19), were astonishing.

> When β particles fall on a plate, a strong radiation emerges from the same side of the plate as that on which the β-particles fall. This radiation is regarded by many observers as a secondary radiation, but more recent experiments seem to show that it consists mainly of primary β-particles, which have been scattered inside the material to such an extent that they emerge again at the same side of

the plate. For α-particles a similar effect has not previously been observed, and is perhaps not to be expected on account of the relatively small scattering which α-particles suffer in penetrating matter.

In the following experiments, however, conclusive evidence was found of the existence of a diffuse reflection of the α-particles. A small fraction of the α-particles falling upon a metal plate have their directions changed to such an extent that they emerge again at the side of incidence. To form an idea of the way in which this effect takes place, the following three points were investigated:—

(I) The relative amount of reflection from different metals.

(II) The relative amount of reflection from a metal of varying thickness.

(III) The fraction of the incident α-particles which are reflected. . . .

On account of the fact that the amount of reflection is very small, it was necessary to use a very intense source of α-rays. A tube was employed similar to that which has been proved to be a suitable source in the scattering experiments of one of us. This source consisted of a glass tube AB (Figure 11), drawn down conically and filled

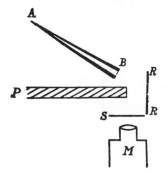

Figure 11

with radium emanation, the end B of the tube being closed airtight by means of a mica window. The thickness of the mica was equivalent to about 1 cm. of air, so that the α-particles could easily pass through it. . . .

The number of α-particles expelled per second through the window was, therefore, very great, and, on account of the small pressure inside the tube . . .

They could see when an α particle struck the zinc sulfide screen, because the α particle caused a minute flash of light or scintillation to appear on the screen.

The zinc sulphide screen S (Figure 11) was fixed behind the lead plate P, in such a position that no α-particles could strike it directly. When a reflector was placed in the position RR at about 1 cm. from the end of the tube, scintillations were at once observed. At the same time the screen brightened up appreciably on account of the reflected β-particles.

By means of a low power microscope,* the number of scintillations per minute on a definite square millimetre of the screen was counted for reflectors of different materials. Care was taken that the different reflectors were always placed in exactly the same position. . . .

In the following table the number of scintillations observed per minute are given in column 3; . . .

The case of lead appears to be an exception which may be due to slight impurities in the lead.

1. Metal.	2. Atomic weight, A.	3. Number of scintillations per minute, Z.	4. A/Z
Lead	207	62	30
Gold	197	67	34
Platinum	195	63	33
Tin	119	34	28
Silver	108	27	25
Copper	64	14·5	28
Iron	56	10·2	18·5
Aluminum	27	3·4	12·5

These results indicated that the number of deflections per minute (as indicated by the number of scintillations on the zinc sulfide screen) increased as elements of higher atomic weight were bombarded. They then investigated the relative amount of deflection (reflection) from various thickness of metals.

(II) We have already pointed out that the diffuse reflection of the α-particles is a consequence of their scattering. According to this point of view, the number of particles reflected must vary with the thickness of the reflecting screen. Since gold can be obtained in very thin and uniform foils, different numbers of these foils were used as reflectors. . . .

The first point on the curve represents the number of scintillations observed for a glass plate alone as reflector; the last point

*(M)

(marked 30) gives the number of scintillations when a thick gold plate was used.

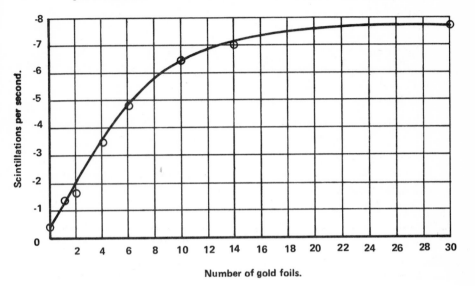

Number of gold foils.

Figure 12.

It brings out clearly that the reflection is not a surface but a volume effect.

Compared, however, with the thickness of gold which an α-particle can penetrate, the effect is confined to a relatively thin layer. In our experiment, about half of the reflected particles were reflected from a layer equivalent to about 2 mm. of air. If the high velocity and mass of the α-particle be taken into account, it seems surprising that some of the α-particles, as the experiment shows, can be turned within a layer of 6×10^{-5} cm. of gold through an angle of 90°, and even more. To produce a similar effect by a magnetic field, the enormous field of 10^9 absolute units would be required.

They then attacked a third problem.

(III) In the next experiment, an estimate of the total number of particles reflected was aimed at. . . .

The arrangement, which is sketched in Figure 13, was such that the α-particles from the plate A fell upon the platinum reflector R, of about 1 square centimetre area, at an average angle of 90°. The reflected particles were counted on different points of the screen S. . . .

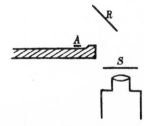

Figure 13.

Three different determinations showed that of the incident α-particles about 1 in 8000 was reflected, under the described conditions.

The results presented in the preceding paper were very difficult to explain with the then generally accepted "plum pudding" model of the atom. Why did most α particles go sailing through gold or platinum foil without being deflected at all, or only deflected two degrees, while an occasional one would be deflected through large angles or even deflected nearly straight backwards? The α particle, remember, is quite massive (atomic weight 4), and travels at high velocities. That an α particle could be bounced back by a thin gold foil so astonished Rutherford, that he described his astonishment with his oft-quoted analogy of firing a 15-inch shell at tissue paper and having it bounce back! Rutherford applied his genius to the interpretation of Geiger and Marsden's work, and in 1911 described a model of the atom which could account for all of the α particle scattering data (38).

In presenting his model, Rutherford defended several assumptions. The scattering of α particles had always been considered the result of many small deflections as the α particles passed among the atoms of the metal foil. Rutherford defended the position that in the wide angle scattering the deflection is due to the encounter of an α particle with a *single* atom. If this is true, how is it possible to account for the incredible force necessary to radically alter the path of a massive and fast moving α particle? Rutherford decided that the forces involved must be electrostatic and invoked Coulomb's Law to explain how the necessarily huge forces could be developed within a single atom. Coulomb's Law states that the forces of interaction between charged objects varies directly with the product of the charges and inversely with the square of the distance between them. Rutherford reasoned that a huge charge must be concentrated in the center of the atom with which a particle would interact if it closely approached the center of the atom. It turns out that you get the same effect whether the concentrated central charge is positive or negative, but Rutherford preferred the assumption that it was positively charged. If all the positive charge, and most of the mass of the atom were concentrated in a central very tiny nucleus, then most α particles would pass through the atom without being affected by it. But an occasional α particle which happened to be heading directly at the nucleus would be bounced back by the tremendous electrostatic force of repulsion developed as the α particle closely approached the concentrated central charge. This model would account for all the scattering patterns of α particles observed by Geiger and Marsden. The word "nucleus" is not used by Rutherford in this first paper (38), but within a year this name was applied to the concentrated central charge and mass. The atom would normally be neutral since sufficient electrons would be located in the remaining space within the atom to balance the charge on the nucleus.

It is well known that the α and β particles suffer deflexions from their rectilinear paths by encounters with atoms of matter. This scattering is far more marked for the β than for the α particle on account of the much smaller momentum and energy of the former particle. There seems to be no doubt that such swiftly moving particles pass through the atoms in their path, and that the deflexions observed are due to the strong electric field traversed within the atomic system. It has generally been supposed that the scattering of a pencil of α or β rays in passing through a thin plate of matter is the result of a multitude of small scatterings by the atoms of matter traversed. The observations, however, of Geiger and Marsden on the scattering of α rays indicate that some of the α particles must suffer a deflexion of more than a right angle at a single encounter. They found, for example, that a small fraction of the incident α particles, about 1 in 20,000, were turned through an average angle of 90° in passing through a layer of gold-foil about ·00004 cm. thick, which was equivalent in stopping-power of the α particle to 1·6 millimetres of air. Geiger showed later that the most probable angle of deflexion for a pencil of α particles traversing a gold-foil of this thickness was about 0°·87. A simple calculation based on the theory of probability shows that the chance of an α particle being deflected through 90° is vanishingly small. In addition, it will be seen later that the distribution of the α particles for various angles of large deflexion does not follow the probability law to be expected if such large deflexions are made up of a large number of small deviations. It seems reasonable to suppose that the deflexion through a large angle is due to a single atomic encounter, for the chance of a second encounter of a kind to produce a large deflexion must in most cases be exceedingly small. A simple calculation shows that the atom must be a seat of an intense electric field in order to produce such a large deflexion at a single encounter. . . .

We shall first examine theoretically the single encounters†
with an °atom of simple structure, which is able to produce large
deflexions of an α particle, and then compare the deductions from
the theory with the experimental data available.

†The deviation of a particle throughout a considerable angle
from an encounter with a single atom will in this paper be called
"single" scattering. The deviation of a particle resulting from a multitude of small deviations will be termed "compound" scattering.

Rutherford now described his model. N is the number of charges in
the central nucleus, and also, since an atom is normally neutral, the
number of oppositely charged units surrounding the nucleus. He assumed that the nucleus is positively charged.

Consider an atom which contains a charge ±Ne at its centre
surrounded by a sphere of electrification containing a charge ∓Ne
supposed uniformly distributed throughout a sphere of radius R. e
is the fundamental unit of charge, which in this paper is taken as
$4 \cdot 65 \times 10^{-10}$ E.S. unit. We shall suppose that for distances less
than 10^{-12} cm. the central charge and also the charge on the α
particle may be supposed to be concentrated at a point. It will be
shown that the main deductions from the theory are independent of
whether the central charge is supposed to be positive or negative.
For convenience, the sign will be assumed to be positive. . . .

In order to form some idea of the forces required to deflect an
α particle through a large angle, consider an atom containing a positive charge Ne at its centre, and surrounded by a distribution of
negative electricity Ne uniformly distributed within a sphere of
radius R. . . .

He then stated that when an α particle approaches the nucleus, the
attractive force of the electrons around the nucleus can be ignored
and the deflection results from the repulsive force of the nucleus.

Suppose an α particle of mass m and velocity v and charge E shot
directly towards the centre of the atom. It will be brought to rest at
a distance b from the centre . . .

Assuming that the central charge is 100 e, it can be calculated
that the value of b for an α particle of velocity $2 \cdot 09 \times 10^9$ cms.
per second is about $3 \cdot 4 \times 10^{-12}$ cm. In this calculation b is supposed to be very small compared with R. Since R is supposed to be
of the order of the radius of the atom, viz. 10^{-8} cm. it is obvious
that the α particle before being turned back penetrates so close to
the central charge, that the field due to the uniform distribution of
negative electricity may be neglected. In general, a simple calculation
shows that for all deflexions greater than a degree, we may without

sensible error suppose the deflexion due to the field of the central charge alone. . . .

Rutherford then showed that an α particle passing close to the nucleus will be deflected in an hyperbolic pathway.

Consider the passage of a positive electrified particle close to the centre of an atom. Supposing that the velocity of the particle is not appreciably changed by its passage through the atom, the path of the particle under the influence of a repulsive force varying inversely as the square of the distance will be an hyperbola with the centre of the atom S as the external focus. . . .

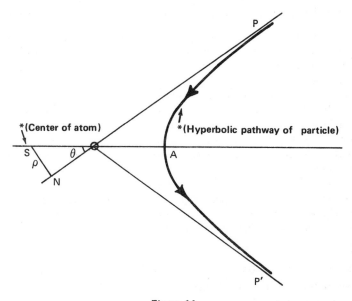

Figure 14.

Rutherford calculated the probability of a single deflection through any particular angle and then continued his discussion.

In these calculations, it is assumed that the α particles scattered through a large angle suffer only one large deflexion. For this to hold, it is essential that the thickness of the scattering material should be so small that the chance of a second encounter involving another large deflexion is very small. . . .

The angular distribution of the α particles scattered from a thin metal sheet affords one of the simplest methods of testing the general correctness of this theory of single scattering. This has been done recently for α rays by Dr. Geiger, who found that the distribution for particles deflected between 30° and 150° from a thin gold-foil was in substantial agreement with the theory. A more detailed account of these and other experiments to test the validity of the theory will be published later. . . .

He then drew some conclusions from the experimental results.

Geiger showed that the most probable angle of deflexion for an atom was nearly proportional to its atomic weight. It consequently follows that the value of N for different atoms should be nearly proportional to their atomic weights, at any rate for atomic weights between gold and aluminium.

Since the atomic weight of platinum is nearly equal to that of gold, it follows from these considerations that the magnitude of the diffuse reflexion of α particles through more than 90° from gold and the magnitude of the average small angle scattering of a pencil of rays in passing through gold-foil are both explained on the hypothesis of single scattering by supposing the atom of gold has a central charge of about $100e$. . . .

Since gold has an atomic weight of 197, the charge on the nucleus would be approximately 1/2 of the atomic weight. This, incidentally would also be the number of electrons in a gold atom if the total atom has a neutral charge.

Rutherford then gave a summary of his results.

General Considerations.

In comparing the theory outlined in this paper with the experimental results, it has been supposed that the atom consists of a central charge supposed concentrated at a point, and that the large single deflexions of the α and β particles are mainly due to their passage through the strong central field. The effect of the equal and opposite compensating charge supposed distributed uniformly throughout a sphere has been neglected. Some of the evidence in support of these assumptions will now be briefly considered. For concreteness, consider the passage of a high speed α particle through an atom having a positive central charge Ne, and surrounded by a compensating charge of N electrons. Remembering that the mass, momentum, and kinetic energy of the α particle are very large compared with the corresponding values for an electron in rapid motion, it does not seem possible from dynamic considerations that an α particle can be deflected through a large angle by a close approach to an electron, even if the latter be in rapid motion and constrained by strong electrical forces. It seems reasonable to suppose that the chance of single deflexions through a large angle due to this cause, if not zero, must be exceedingly small compared with that due to the central charge.

It is of interest to examine how far the experimental evidence throws light on the question of the extent of the distribution of the central charge. . . .

Considering the evidence as a whole, it seems simplest to suppose that the atom contains a central charge distributed through a very small volume, and that the large single deflexions are due to the central charge as a whole, and not to its constituents. At the same time, the experimental evidence is not precise enough to negative the possibility that a small fraction of the positive charge may be carried by satellites extending some distance from the centre. . . .

The general data available indicate that the value of this central charge for different atoms is approximately proportional to their atomic weights, at any rate for atoms heavier than aluminium. It will be of great interest to examine experimentally whether such a simple relation holds also for the lighter atoms. . . .

Rutherford, after referring to Nagaoka's Saturnian model, speculated that the α particle (or helium atom) might be the fundamental building block of the atom. Similar speculation was to continue for many years.

It is of interest to note that Nagaoka has mathematically considered the properties of a "Saturnian" atom which he supposed to consist of a central attracting mass surrounded by rings of rotating electrons. He showed that such a system was stable if the attractive force was large. From the point of view considered in this paper, the chance of large deflexion would practically be unaltered, whether the atom is considered to be a disk or a sphere. It may be remarked that the approximate value found for the central charge of the atom of gold ($100e$) is about that to be expected if the atom of gold consisted of 49 atoms of helium, each carrying a charge $2\ e$. This may be only a coincidence, but it is certainly suggestive in view of the expulsion of helium atoms carrying two unit charges from radioactive matter.

Rutherford hedged a little, and pointed out that the experimental results could equally well be explained by a negatively charged nucleus. However, throughout the paper he had assumed a positive central charge which seems to indicate his position on the subject.

The deductions from the theory so far considered are independent of the sign of the central charge, and it has not so far been found possible to obtain definite evidence to determine whether it be positive or negative. It may be possible to settle the question of sign by consideration of the difference of the laws of absorption of the β particle to be expected on the two hypotheses, for the effect of radiation in reducing the velocity of the β particle should be far more marked with a positive than with a negative centre. If the central charge be positive, it is easily seen that a positively charged mass if released from the centre of a heavy atom, would acquire a great velocity in moving through the electric field. It may be possible in this way to account for the high velocity of expulsion of α particles without supposing that they are initially in rapid motion within the atom.

Further consideration of the application of this theory to these and other questions will be reserved for a later paper, when the main deductions of the theory have been tested experimentally. Experiments in this direction are already in progress by Geiger and Marsden.

The experiments mentioned at the end of the preceding paper were published two years later (20). These results overwhelmingly supported Rutherford's nuclear model of the atom.

In a former paper one of us has shown that in the passage of α particles through matter the deflexions are, on the average, small and of the order of a few degrees only. In the experiments a narrow pencil of α particles fell on a zinc-sulphide screen in vacuum, and the distribution of the scintillations on the screen was observed when different metal foils were placed in the path of the α particles. From the distribution obtained, the most probable angle of scattering could be deduced, and it was shown that the results could be explained on the assumption that the deflexion of a single α particle is the resultant of a large number of very small deflexions caused by the passage of the α particle through the successive individual atoms of the scattering substance.

In an earlier paper however, we pointed out that α particles are sometimes turned through very large angles. This was made evident by the fact that when α particles fall on a metal plate, a small fraction of them, about 1/8000 in the case of platinum, appears to be diffusely reflected. This amount of reflexion, although small, is, however, too large to be explained on the above simple theory of scattering. It is easy to calculate from the experimental data that the probability of a deflexion through an angle of 90° is vanishingly small, and of a different order to the value found experimentally.

Professor Rutherford has recently developed a theory to account for the scattering of α particles through these large angles, the assumption being that the deflexions are the result of an intimate encounter of an α particle with a single atom of the matter traversed. In this theory an atom is supposed to consist of a strong positive or negative central charge concentrated within a sphere of less than about 3×10^{-12} cm. radius, and surrounded by electricity of the opposite sign distributed throughout the remainder of the atom of about 10^{-8} cm. radius. In considering the deflexion of an α particle directed against such an atom, the main deflexion-effect can be supposed to be due to the central concentrated charge which will cause the α particle to describe an hyperbola with the centre of the atom as one focus. . . .

The fraction of the α particles whose paths are sufficiently near to the centre of the atom will, however, be small so that the probability of an α particle suffering a large deflexion of this nature will be correspondingly small. . . .

Geiger and Marsden then presented their experimental results. Their comments indicate the degree to which the experimental results support the nuclear model. Rarely in a scientific paper, where for the most part absolutes are avoided like the plague, do you encounter a phrase like "completely verified."

In the previous sections we have completely verified the theory given by Prof. Rutherford. Since according to this theory, the large deflexion of an α particle is the result of a close encounter with a single atom of matter, it is possible to calculate the magnitude of the central charge of the atom when the fraction of α particles scattered under definite conditions is determined. We have made several attempts under different conditions to obtain a quantitative estimate of the scattered particles, but the results so far have only given us an approximate value. . . .

The results showed that, using a gold foil of air equivalent 1 mm. . . .

scattered through an angle of 45° and observed on an area of 1 sq. mm. placed normally at a distance of 1 cm. from the point of incidence of the beam was $3 \cdot 7 \times 10^{-7}$. Substituting this value in the equation* given at the commencement of this paper, it can be calculated that the value of the number of elementary electric charges composing the central charge of the gold atom is about half the atomic weight. This result is probably correct to 20 per cent., and agrees with the deduction of Prof. Rutherford from the less definite data given in our previous paper. . . .

They then carefully distinguished between the single encounters of α particles as discussed by Rutherford, and the multiple small scatterings produced when an α particle traverses a great number of atoms.

*(not reproduced here)

It will be seen that the laws of "single scattering" found in this paper are quite distinct from the laws of "compound scattering" previously deduced by Geiger. It must be remembered, however, that the experiments are not directly comparable. In the present paper we are dealing with very thin sheets of matter, and are measuring the very small fraction of α particles which are deflected by single collisions through relatively large angles. The experiments of Geiger, however, deal with larger thicknesses of scattering foils and angles of deflexion of a few degrees only. Under these conditions the scattering is due to the combination of a large number of deflexions not only by the central charges of the atoms, but probably also by the electronic charge distributed throughout the remainder of their volumes.

The summary given below is important because it shows how completely the experimental results supported Rutherford's nuclear model of the atom.

Summary.

The experiments described in the foregoing paper were carried out to test a theory of the atom proposed by Prof. Rutherford, the main feature of which is that there exists at the centre of the atom an intense highly concentrated electrical charge. The verification is based on the laws of scattering which were deduced from this theory. The following relations have been verified experimentally:—

(1) The number of α particles emerging from a scattering foil at an angle ϕ with the original beam varies as $1/\sin^4 \phi/2$, when the α particles are counted on a definite area at a constant distance from the foil. This relation has been tested for angles varying from $5°$ to $150°$, and over this range the number of α particles varied from 1 to 250,000 in good agreement with the theory.

(2) The number of α particles scattered in a definite direction is directly proportional to the thickness of the scattering foil for small thicknesses. For larger thicknesses the decrease of velocity of the α particles in the foil causes a somewhat more rapid increase in the amount of scattering.

(3) The scattering per atom of foils of different materials varies approximately as the square of the atomic weight. This relation was tested for foils of atomic weight from that of carbon to that of gold.

(4) The amount of scattering by a given foil is approximately proportional to the inverse fourth power of the velocity of the incident α particles. This relation was tested over a range of velocities such that the number of scattered particles varied as $1:10$.

(5) Quantitative experiments show that the fraction of a α particles of Ra C, which is scattered through an angle of $45°$ by a

gold foil of 1 mm. air equivalent ($2 \cdot 1 \times 10^{-5}$ cm.), is $3 \cdot 7 \times 10^{-7}$ when the scattered particles are counted on a screen of 1 sq. mm. area placed at a distance of 1 cm. from the scattering foil. From this figure and the foregoing results, it can be calculated that the number of elementary charges composing the centre of the atom is equal to half the atomic weight.

The following year Rutherford presented a summation of the evidence supporting his nuclear model of the atom (40). It was presented in Rutherford's typically brilliant, clear, straightforward prose. It is an excellent summary of the material presented so far in this chapter.

In my previous paper . . . I pointed out the importance of the study of the passage of the high speed α and β particles through matter as a means of throwing light on the internal structure of the atom. Attention was drawn to the remarkable fact, first observed by Geiger and Marsden, that a small fraction of the swift α particles from radioactive substances were able to be deflected through an angle of more than $90°$ as the results of an encounter with a single atom. It was shown that the type of atom devised by Lord Kelvin and worked out in great detail by Sir J.J. Thomson* was unable to produce such large deflexions unless the diameter of the positive sphere was exceedingly small. In order to account for this large angle scattering of α particles, I supposed that the atom consisted of a positively charged nucleus of small dimensions in which practically all the mass of the atom was concentrated. The nucleus was supposed to be surrounded by a distribution of electrons to make the atom electrically neutral, and extending to distances from the nucleus comparable with the ordinary accepted radius of the atom. Some of the swift α particles passed through the atoms in their path and entered the intense electric field in the neighbourhood of the nucleus and were deflected from their rectilinear path. In order to suffer a deflexion of more than a few degrees, the α particle has to pass very close to the nucleus, and it was assumed that the field of force in this region was not appreciably affected by the external electronic distribution. Supposing that the forces between the nucleus and the α particle are repulsive and follow the law of inverse squares, the α particle describes a hyperbolic orbit round the nucleus and its deflexion can be simply calculated. . . .

From the data of scattering on α particles previously given by Geiger, it was deduced that the value of the nucleus charge was equal

*(plum pudding model)

to about half the atomic weight multiplied by the electronic charge. . . .

The experimental results of Geiger and Marsden were thus in complete accord with the predictions of the theory, and indicated the essential correctness of this hypothesis of the structure of the atom. . . .

It is of interest to note that C.T.R. Wilson, by photographing the trails of the α particle, later showed that the α particle occasionally suffers a sudden deflexion through a large angle. This affords convincing evidence of the correctness of the view that large deflexions do occasionally occur as a result of an encounter with a single atom. . . .

In fact, the passage of swift α particles through matter affords the most definite and straightforward method of throwing light on the gross structure of the atom, for the α particle is able to penetrate the atom without serious disturbance from the electronic distribution, and thus is only affected by the intense field associated with the nucleus of the atom.

Two important problems introduced or left unresolved by Rutherford's nuclear model of the atom were: (1) to determine the exact charge on the nucleus of the atom, and (2) to determine the configuration of the electrons around the nucleus. The latter problem was particularly difficult since electrons moving around the nucleus would, according to classical physics, produce an unstable atom.

Both of these problems were attacked and solved by examining the interaction of matter and energy. Energy can be transferred from one matter system to another in the form of electromagnetic radiation. It had been established by 1912 that this radiant energy seemed to possess dual properties. In some respects it acts as a wave phenomenon, having for example, properties we associate with moving waves of water. In other situations, the radiant energy acts as though it were a stream of particles, the packets of energy (called quanta) having properties analogous to particles of matter.

In determining the charge on the nucleus, we will be concerned with the wave properties of radiant energy. We have referred to energy as "electromagnetic" because it is transmitted through space as though it were composed of a combination of electrical and magnetic waves at right angles to one another. A wave is generally described as a series of crests and troughs. The distance between succeeding crests is called the wave length and the number of crests passing a point in a specific time, the frequency of the wave. So it is in our description of electromagnetic radiation. All electromagnetic radiation travels at the same velocity in a vacuum, 186,000 miles per second, and forms a continuous spectrum from very long wave length, low frequency radio waves through the very short wave length, high frequency γ rays. Intermediate wave lengths form successively, from low frequency to high, the infrared, visible light, ultraviolet and X-ray parts of the spectrum. The higher the frequency of a wave, the more energy it possesses.

First, let us direct our attention to the X-ray portion of the spectrum. By 1913, it was already suspected that when in a cathode ray tube cathode rays strike the atoms of the anode, the resulting vibration of the atom's innermost electrons radiates out energy in the X-ray end of the spectrum. A new series of techniques, in which X-rays were reflected by crystals, made it possible to determine the frequencies of the X-rays produced by different elements as they were bombarded by cathode rays.

Following several brief introductory passages by the Dutch physicist Van den Broek, the brilliant work of Moseley concerning the X-rays emitted by different elements is presented in detail. This work solved the problem of determining exact charge on the nucleus of the atom.

Rutherford and his co-workers had estimated that the positive charge on the nucleus was equal to half of the atomic weight, but admitted a possible variation up to 20%. Van den Broek in 1913, suggested that the charge on the nucleus was the same as the number of the element on the periodic table. In other words, the positive charge on the nucleus of hydrogen would be 1, on helium 2, on lithium 3, *etc.* First, he suggests (63) that the number of possible elements is equal to the number of positive charges in the nucleus (which will be the same as the number of electrons outside the nucleus for a neutral atom).

> Hence, if this* cubic periodic system should prove to be correct, then the number of possible elements is equal to the number of possible permanent charges of each sign per atom, or to each possible permanent charge (of both signs) per atom belongs a possible element.

*(Mendeleev's)

He expanded on this idea later the same year, beginning with quotes from some of his earlier references. He then pointed out that if you use the number of an element's position in the periodic table instead of its atomic weight, certain of Geiger and Marsden's experimental data are easier to explain (64).

> In a previous letter to Nature (July 20, 1911, p. 78) the hypothesis was proposed that the atomic weight being equal to about twice the intra-atomic charge, "to each possible intra-atomic charge corresponds a possible element," or that (*Phys. Zeitschr.*, xiv., 1912, p. 39), "if all elements be arranged in order of increasing atomic weights, the number of each element in that series must be equal to its intra-atomic charge."
>
> Charges being known only very roughly (probably correct to 20 per cent.), and the number of the last element Ur in the series not being equal even approximately to half its atomic weight, either the number of elements in Mendeléeff's system is not correct (that was supposed to be the case in the first letter), or the intra-atomic charge for the elements at the end of the series is much smaller than that deduced from experiment (about 100 for Au).
>
> Now, according to Rutherford, the ratio of the scattering of α particles per atom divided by the square of the charge must be constant. Geiger and Marsden . . .
>
> putting the nuclear charge proportional to the atomic weight, found values, however, showing, not constancy, but systematic deviation from (mean values) 3·825 for Cu to 3·25 for Au. If now in these values the number M of the place each element occupies in Mendeléeff's series is taken instead of A, the atomic weight, we get a real constant (18·7±0·3); hence the hypothesis proposed holds good for Mendeléeff's series, but the nuclear charge is not equal to half the atomic weight.

During this time Moseley was completing his work on the frequencies of X-rays emitted by elements bombarded by cathode rays. His results provided convincing quantitative proof of Van den Broek's ideas. His first paper was published in late 1913, and a second the following April. The first paper has become a classic (27).

> In the absence of any available method of spectrum analysis, the characteristic types of X radiation, which an atom emits when suitably excited, have hitherto been described in terms of their absorption in aluminium. The interference phenomena exhibited by X rays when scattered by a crystal have now, however, made possible the accurate determination of the frequencies of the various types of radiation. . . .

The present paper contains a description of a method of photographing these spectra, which makes the analysis of the X rays as simple as any other branch of spectroscopy. . . .

The results already obtained show that such data have an important bearing on the question of the internal structure of the atom and strongly support the views of Rutherford and of Bohr.

. . . an element excited by a stream of sufficiently fast cathode rays emits its characteristic X radiation. He used as targets a number of substances mounted on a truck inside an exhausted tube. A magnetic device enabled each target to be brought in turn into the line of fire. This apparatus was modified to suit the present work. . . .

Twelve elements have so far been examined. The ten given in Table I. were chosen as forming a continuous series with only one gap. It was hoped in this way to bring out clearly any systematic results. The inclusion of nickel was of special interest owing to its anomalous position in the periodic system. Radiations from these substances are readily excited, and the large angles of reflexion make it easy to measure the wave lengths with accuracy.

Element.	$Q = (\nu / \frac{3}{4} \nu_0)^{1/2}$	Natomic Number.	Atomic Weight.
CALCIUM.	19·00	20	40·09
SCANDIUM.	. . .	21	44·1
TITANIUM.	20·99	22	48·1
VANADIUM.	21·96	23	51·06
CHROMIUM.	22·98	24	52·0
MANGANESE.	23·99	25	54·93
IRON.	24·90	26	55·85
COBALT.	26·00	27	58·97
NICKEL.	27·04	28	58·68
COPPER.	28·01	29	63·57
ZINC.	20·01	30	65·37

In the last column of the table, all the elements but cobalt and nickel are listed by increasing atomic weight. These last two appear in reverse order since cobalt's atomic weight is greater than nickel's. They are listed in the proper order however, when grouped by chemical properties. This is one of the anomalies in Mendeleev's periodic table which disappears with the work of Moseley! The number of positive charges in the nucleus, not the atomic weight controls the chemical properties of the elements. This number of positive charges Moseley calls the atomic number.

He then described the plate in which the photographs of the two principal X-ray frequencies emitted by each element are presented.

> Plate XXIII. shows the spectra in the third order placed approximately in register. Those parts of the photographs which represent the same angle of reflexion are in the same vertical line. The actual angles can be taken from Table I. It is to be seen that the spectrum of each element consists of two lines. Of these the stronger has been called α in the table, and the weaker β. The lines found on any of the plates besides α and β were almost certainly all due to impurities. . . .

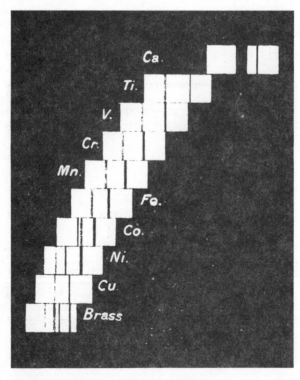

The prevalence of lines due to impurities suggests that this may prove a powerful method of chemical analysis. Its advantage over ordinary spectroscopic methods lies in the simplicity of the spectra and the impossibility of one substance masking the radiation from another. It may even lead to the discovery of missing elements, as it will be possible to predict the position of their characteristic lines. . .

A discussion will now be given of the meaning of the wavelengths* found for the principal spectrum line α. . . .

In this discussion Moseley showed that each successive element presented in the table shows an increase in *equal* increments of the quantity Q (which is a measure of the frequency of the X-rays emitted). This must be a measure of the charge on the nucleus. Since the charge on the nucleus of each successive element is larger, the frequency of X-rays emitted by the vibrating inner electrons should increase, due to the greater pull on the inner electrons by the nucleus.

The reason for introducing this particular constant will be given later. It is at once evident that Q increases by a constant amount as we pass from one element to the next, using the chemical order of the elements in the periodic system. Except in the case of nickel and cobalt, this is also the order of the atomic weights. While, however, Q increases uniformly the atomic weights vary in an apparently arbitrary manner, so that an exception in their order does not come as a surprise. We have here a proof that there is in the atom a fundamental quantity, which increases by regular steps as we pass from one element to the next. This quantity can only be the charge on the central positive nucleus, of the existence of which we already have definite proof. Rutherford has shown, from the magnitude of the scattering of α particles by matter, that this nucleus carries a + charge approximately equal to that of $\frac{A}{2}$ electrons, where A is the atomic weight. Barkla, from the scattering of X rays by matter, has shown that the number of electrons in an atom is roughly $\frac{A}{2}$, which for an electrically neutral atom comes to the same thing. Now atomic weights increase on the average by about 2 units at a time, and this strongly suggests the view that N increases from atom to atom always by a single electronic unit. We are therefore led by experiment to the view that N is the same as the number of the place occupied by the element in the periodic system. This atomic number is then for H 1 for He 2 for Li 3 . . . for Ca 20 . . . for Zn 30, &c. This theory was originated by Broek and since used by Bohr. We can

*(X-ray)

confidently predict that in the few cases in which the order of the atomic weights A clashes with the chemical order of the periodic system, the chemical properties are governed by N; while A is itself probably a complicated function of N. The very close similarity between the X-ray spectra of the different elements shows that these radiations originate inside the atom, and have no direct connexion with the complicated light-spectra and chemical properties which are governed by the structure of its surface.

In a second paper (28) Moseley increased the number of elements considered, and then commented on the significance of his results.

The first part of this paper dealt with a method of photographing X-ray spectra, and included the spectra of a dozen elements. More than thirty other elements have now been investigated, and simple laws have been found which govern the results, and make it possible to predict with confidence the position of the principal lines in the spectrum of any element from aluminium to gold. . . .

Now Rutherford has proved that the most important constituent of an atom is its central positively charged nucleus, and Van den Broek has put forward the view that the charge carried by this nucleus is in all cases an integral multiple of the charge on the hydrogen nucleus. There is every reason to suppose that the integer which controls the X-ray spectrum is the same as the number of electrical units in the nucleus, and these experiments therefore give the strongest possible support to the hypothesis of Van den Broek.

Moseley then summarized his results.

Summary.

1. Every element from aluminum to gold is characterized by an integer N which determines its X-ray spectrum. Every detail in the spectrum of an element can therefore be predicted from the spectra of its neighbours.
2. This integer N, the atomic number of the element, is identified with the number of positive units of electricity contained in the atomic nucleus.
3. The atomic numbers for all elements from Al to Au have been tabulated on the assumption that N for Al is 13.
4. The order of the atomic numbers is the same as that of the atomic weights, except where the latter disagrees with the order of the chemical properties.
5. Known elements correspond with all the numbers between 13 and 79 except three. There are here three possible elements still undiscovered.

The importance of Moseley's findings is hard to overestimate. Some indication of its importance can be seen in a statement by Rutherford (39) published during the same month as Moseley's first paper.

> The original suggestion of van der Broek that the charge on the nucleus is equal to the atomic number and not to half the atomic weight seems to me very promising. This idea has already been used by Bohr in his theory of the constitution of atoms. The strongest and most convincing evidence in support of this hypothesis will be found in a paper by Moseley in *The Philosophical Magazine* of this month. He there shows that the frequency of the X radiations from a number of elements can be simply explained if the number of unit charges on the nucleus is equal to the atomic number. It would appear that the charge on the nucleus is the fundamental constant which determines the physical and chemical properties of the atom, while the atomic weight, although it approximately follows the order of the nucleus charge, is probably a complicated function of the latter depending on the detailed structure of the nucleus.

In retrospect, Rutherford again discussed the importance of Moseley's work, this time in a 1920 lecture (45).

> The importance of the atomic number in fixing the properties of an atom was shown by the remarkable work of Moseley on the X-ray spectra of the elements. He showed that the frequency of vibration of corresponding lines in the X-ray spectra of the elements depended on the square of a number which varied by unity in successive elements. This relation received an interpretation by supposing that the nuclear charge varied by unity in passing from atom to atom, and was given numerically by the atomic number. I can only emphasize in passing the great importance of Moseley's work, not only in fixing the number of possible elements, and the position of undetermined elements, but in showing that the properties of an atom were defined by a number which varied by unity in successive atoms. This gives a new method of regarding the periodic classification of the elements, for the atomic number, or its equivalent the nuclear charge, is of more fundamental importance than its atomic weight.

Another series of papers, critically important in the history of the atom, were also published in 1913. These were written by the Danish physicist Niels Bohr, who for part of a year worked with Rutherford in his Manchester, England laboratory. He became interested in attacking the second major problem with Rutherford's nuclear model, the configuration of the electrons around the nucleus. Bohr's solution to the problem involved electromagnetic radiation acting as "quanta" or particle-like packages of energy.

His work involved principally the visible or light range of the electromagnetic spectrum. It was well established that atoms in the gaseous state, heated to high temperatures, could absorb certain particular frequencies of light and remit those same frequencies. Atoms of each different element can absorb and emit only specific frequencies (in Chapter 3 these were compared to fingerprints because they can be used to identify the elements). If the light emitted by atoms is directed through a prism, the individual frequencies will be separated and appear as a series of bright lines, called the bright-line spectrum. Rutherford's nuclear model, as we have examined it so far, cannot explain why specific frequencies are absorbed and emitted by the atoms of an element. A great deal of evidence had been amassed which pointed to the electrons as the absorbers and emitters of those specific frequencies in the light (and ultraviolet) range. Hence, Bohr directed his attention to the electron configuration of the atom. Several problems were immediately apparent. If the electrons were not moving, then the attractive forces of the positively-charged nucleus would simply pull the much lighter, negatively-charged electrons into the nucleus causing the atom to collapse. Rutherford had assumed, therefore, that the electrons moved in orbits around the nucleus. This however, brought up another problem. Classical physics demonstrated that any charged particle which is accelerated radiates away energy. An electron in orbit around the nucleus would undergo a constant acceleration (change in direction from a straight line path). The electron should therefore lose energy, slow down, and gradually spiral into the nucleus, again causing the atom to collapse.

Bohr constructed a model of the hydrogen atom, the simplest of all atoms with the single positive charge on its nucleus and the single electron forming a neutral atom. In order to explain the motion of the electron, Bohr was forced to use Plank's mathematical description of energy emitted from an atom. Plank had assumed that energy was not continuously emitted from matter, but was released in discrete packages called quanta (thus giving light particle-like characteristics). His basic equation was a very simple one relating the amount of energy emitted in a quantum and the frequency of light this would produce.

Bohr had to make some very wild assumptions. He assumed that there are certain specific orbits in which electrons can move *without* radiating away energy. This assumption was contrary to classical physics, but unless he made this assumption he could not explain the behavior of the electron. He further assumed that the electron could absorb a particular quantum of energy and leap to an orbit further from the nucleus. When the electron returned to its original orbit it would re-emit this same quantum of energy. If there were many possible orbits, then many quanta representing different frequencies of light could be absorbed or emitted. This could account for the bright line spectrum of the atom. Each bright emission line would represent the electron's leap from one specific orbit to another. Bohr made other assumptions which defined the specific orbits for the electron and enabled him to calculate the frequencies of the bright line spectrum of hydrogen. His predicted frequencies were very close to the experimentally observed ones!

He began his first paper of 1913 (8) with a description of the nuclear model of the atom.

In order to explain the results of experiments on scattering of α rays by matter Prof. Rutherford has given a theory of the structure of atoms. According to this theory, the atoms consist of a positively charged nucleus surrounded by a system of electrons kept together by attractive forces from the nucleus; the total negative charge of the electrons is equal to the positive charge of the nucleus. Further, the nucleus is assumed to be the seat of the essential part of the mass of the atom, and to have linear dimensions exceedingly small compared with the linear dimensions of the whole atom. The number of electrons in an atom is deduced to be approximately equal to half the atomic weight. Great interest is to be attributed to this atom-model; for, as Rutherford has shown, the assumption of the existence of nuclei, as those in question, seems to be necessary in order to account for the results of the experiments on large angle scattering of the α rays.

In an attempt to explain some of the properties of matter on the basis of this atom-model we meet, however, with difficulties of a serious nature arising from the apparent instability of the system of electrons: difficulties purposely avoided in atom-models previously considered, for instance, in the one proposed by Sir J.J. Thomson. According to the theory of the latter the atom consists of a sphere of uniform positive electrification, inside which the electrons move in circular orbits.

The principal difference between the atom-models proposed

by Thomson and Rutherford consists in the circumstance that the forces acting on the electrons in the atom-model of Thomson allow of certain configurations and motions of the electrons for which the system is in a stable equilibrium; such configurations, however, apparently do not exist for the second atom-model. . . .

Bohr now stated that the difficulties might lie with the classical physics, and not with the nuclear model of the atom. He then introduced Plank's quantum theory.

The way of considering a problem of this kind has, however, undergone essential alterations in recent years owing to the development of the theory of the energy radiation, and the direct affirmation of the new assumptions introduced in this theory, found by experiments on very different phenomena such as specific heats, photoelectric effect, Röntgen-rays, &c. The result of the discussion of these questions seems to be a general acknowledgment of the inadequacy of the classical electrodynamics in describing the behaviour of systems of atomic size. Whatever the alteration in the laws of motion of the electrons may be, it seems necessary to introduce in the laws in question a quantity foreign to the classical electrodynamics, *i.e.*, Planck's constant, or as it often is called the elementary quantum of action. By the introduction of this quantity the question of the stable configuration of the electrons in the atoms is essentially changed, . . .

This paper is an attempt to show that the application of the above ideas to Rutherford's atom-model affords a basis for a theory of the constitution of atoms. It will further be shown that from this theory we are led to a theory of the constitution of molecules.

In the present first part of the paper the mechanism of the binding of electrons by a positive nucleus is discussed in relation to Planck's theory. It will be shown that it is possible from the point of view taken to account in a simple way for the law of the line spectrum of hydrogen. . . .

Bohr demonstrated the inadequacies of the classical physics, and introduced Plank's theory.

Part I.—Binding of Electrons by Positive Nuclei.
General Considerations.

The inadequacy of the classical electrodynamics in accounting for the properties of atoms from an atom-model as Rutherford's, will appear very clearly if we consider a simple system consisting of a positively charged nucleus of very small dimensions and an electron describing closed orbits around it. . . .

Let us at first assume that there is no energy radiation. In this case the electron will describe stationary elliptical orbits. . . .

Let us now, however, take the effect of the energy radiation into account, calculated in the ordinary way from the acceleration of the electron. In this case the electron will no longer describe stationary orbits. . . .

The electron will approach the nucleus describing orbits of smaller and smaller dimensions, and with greater and greater frequency; the electron on the average gaining in kinetic energy at the same time as the whole system loses energy. This process will go on until the dimensions of the orbit are of the same order of magnitude as the dimensions of the electron or those of the nucleus. A simple calculation shows that the energy radiated out during the process considered will be enormously great compared with that radiated out by ordinary molecular processes.

It is obvious that the behaviour of such a system will be very different from that of an atomic system occurring in nature. In the first place, the actual atoms in their permanent state seem to have absolutely fixed dimensions and frequencies. Further, if we consider any molecular process, the result seems always to be that after a certain amount of energy characteristic for the systems in question is radiated out, the systems will again settle down in a stable state of equilibrium, in which the distances apart of the particles are of the same order of magnitude as before the process.

Now the essential point in Planck's theory of radiation is that the energy radiation from an atomic system does not take place in the continuous way assumed in the ordinary electrodynamics, but that it, on the contrary, takes place in distinctly separated emissions, . . .

He then presented his assumptions and calculated the various possible stationary states or orbits of the electron.

Returning to the simple case of an electron and a positive nucleus considered above, let us assume that the electron at the beginning of the interaction with the nucleus was at a great distance apart from the nucleus, and had no sensible velocity relative to the latter. Let us further assume that the electron after the interaction has taken place has settled down in a stationary orbit around the nucleus. We shall, for reasons referred to later, assume that the orbit in question is circular: . . .

Let us now assume that, during the binding of the electron, a homogeneous radiation is emitted of a frequency ν, equal to half the frequency of revolution of the electron in its final orbit . . .

Bohr presented his calculations and concluded:

> If in these expressions we ... *   values, we get a series of
> values ... corresponding to a series of configurations of the system.
> According to the above considerations, we are led to assume that
> these configurations will correspond to states of the system in which
> there is no radiation of energy; states which consequently will be
> stationary as long as the system is not disturbed from outside. . . .
>
> This case** will therefore correspond to the most stable state of
> the system, *i.e.,* will correspond to the binding of the electron for
> the breaking up of which the greatest amount of energy is re-
> quired. . . .

Later he again stated his assumptions.

> Before proceeding it may be useful to restate briefly the
> ideas characterizing the calculations ... The principal assumptions
> used are:
> (1)   That the dynamical equilibrium of the systems in the
>   stationary states can be discussed by help of the ordi-
>   nary mechanics, while the passing of the systems be-
>   tween different stationary states cannot be treated on
>   that basis.
> (2)   That the latter process is followed by the emission of a
>   *homogeneous* radiation, for which the relation between
>   the frequency and the amount of energy emitted is the
>   one given by Planck's theory.
>
> The first assumption seems to present itself; for it is known
> that the ordinary mechanics cannot have an absolute validity, but
> will only hold in calculations of certain mean values of the motion
> of the electrons. . . .
>
> The second assumption is in obvious contrast to the
> ordinary ideas of electrodynamics, but appears to be necessary in
> order to account for experimental facts. . . .

Bohr then used his calculations to predict the spectral lines which
hydrogen should emit.

Emission of Line-spectra.

Spectrum of Hydrogen.—General evidence indicates that an
atom of hydrogen consists simply of a single electron rotating round
a positive nucleus of charge *e*. The reformation of a hydrogen atom,
when the electron has been removed to great distances away from

*(use any small whole number)
**(if 1 is the number used)

> the nucleus—*e.g.*, by the effect of electrical discharge in a vacuum
> tube—will accordingly correspond to the binding of an electron by a
> positive nucleus considered* ...
>
> we get for the total amount of energy radiated out by the formation
> of one of the stationary states ...

After presenting his calculations Bohr summed up his results. The
various series to which he refers are particular sets of frequencies
emitted by hydrogen. τ stands for a small whole number. The
important thing to notice is the high correlation between his predictions and the observed spectral lines.

> We see that this expression accounts for the law connecting
> the lines in the spectrum of hydrogen. If we put $\tau_2 = 2$ and let τ_1
> vary, we get the ordinary Balmer series. If we put $\tau_2 = 3$, we get the
> series in the ultra-red observed by Paschen and previously suspected
> by Ritz. If we put $\tau_2 = 1$ and $\tau_2 = 4,5,..$, we get series respec-
> tively in the extreme ultra-violet and the extreme ultra-red, which
> are not observed, but the existence of which may be expected.
>
> The agreement in question is quantitative as well as qualita-
> tive. . . .

He then demonstrated the predictive value of his model by showing
that some of the spectral lines attributed to hydrogen are actually
emitted by helium!

> It will be observed that we in the above way do not obtain
> other series of lines, generally ascribed to hydrogen; for instance, the
> series first observed by Pickering in the spectrum of the star ζ
> Puppis, and the set of series recently found by Fowler by experi-
> ments with vacuum tubes containing a mixture of hydrogen and
> helium. We shall, however, see that, by help of the above theory, we
> can account naturally for these series of lines if we ascribe them to
> helium. . . .

Bohr commented that the atom remains, most of the time, in its
lowest energy state, with the electron in its lowest energy, most
stable orbit.

> The great number of different stationary states we do not
> observe except by investigation of the emission and absorption of
> radiation. In most of the other physical phenomena, however, we
> only observe the atoms of the matter in a single distinct state, *i.e.*,

*(previously)

the state of the atoms at low temperature.* From the preceding considerations we are immediately led to the assumption that the "permanent" state is the one among the stationary states during the formation of which the greatest amount of energy is emitted. . . .

Then he explained the absorption of energy.

Absorption of Radiation.

In order to account** . . . it is necessary to introduce assumptions on the mechanism of absorption of radiation which correspond to those we have used considering the emission. Thus we must assume that a system consisting of a nucleus and an electron rotating round it under certain circumstances can absorb a radiation of a frequency equal to the frequency of the homogeneous radiation emitted during the passing of the system between different stationary states. . . .

These considerations seem to be in conformity with experiments on absorption in gases. In hydrogen gas at ordinary conditions for instance there is no absorption of a radiation of a frequency corresponding to the line-spectrum of this gas; such an absorption is only observed in hydrogen gas in a luminous state.

Although Bohr had made many radical assumptions, his ability to explain observational facts made his assumptions difficult to refute. In a second paper (9) Bohr attempted to extend his theory to more complex atoms having more than a single positive charge on their nucleus, and hence, more orbiting electrons.

His opening paragraphs are particularly interesting for they show his deep insight into the nuclear model of the atom. This paper was published before Moseley's (page 88), and yet, he already understood and accepted the importance of atomic number.

As in the previous paper, we shall assume that the cluster of electrons is formed by the successive binding by the nucleus of electrons initially nearly at rest, energy at the same time being radiated away. This will go on until, when the total negative charge on the bound electrons is numerically equal to the positive charge on the nucleus, the system will be neutral and no longer able to exert sensible forces on electrons at distances from the nucleus great in comparison with the dimensions of the orbits of the bound electrons. We may regard the formation of helium from α rays as an observed example of a process of this kind, an α particle on this view being identical with the nucleus of a helium atom.

*(room temperature)
**(for certain observations)

On account of the small dimensions of the nucleus, its internal structure will not be of sensible influence on the constitution of the cluster of electrons, and consequently will have no effect on the ordinary physical and chemical properties of the atom. The latter properties on this theory will depend entirely on the total charge and mass of the nucleus; the internal structure of the nucleus will be of influence only on the phenomena of radioactivity.

From the result of experiments on large-angle scattering of α-rays, Rutherford found an electric charge on the nucleus corresponding per atom to a number of electrons approximately equal to half the atomic weight. This result seems to be in agreement with the number of electrons per atom calculated from experiments on scattering of Röntgen radiation. The total experimental evidence supports the hypothesis that the actual number of electrons in a neutral atom with a few exceptions is equal to the number which indicates the position of the corresponding element in the series of elements arranged in order of increasing atomic weight. For example on this view, the atom of oxygen which is the eighth element of the series has eight electrons and a nucleus carrying eight unit charges. . . .

Constitution of Atoms containing very few Electrons.

As stated in §1, the condition of the universal constancy of the angular momentum of the electrons, together with the condition of stability, is in most cases not sufficient to determine completely the constitution of the system. On the general view of formation of atoms, however, and by making use of the knowledge of the properties of the corresponding elements, it will be attempted, in this section and the next, to obtain indications of what configurations of the electrons may be expected to occur in the atoms. In these considerations we shall assume that the number of electrons in the atom is equal to the number which indicates the position of the corresponding element in the series of elements arranged in order of increasing atomic weight. Exceptions to this rule will be supposed to occur only at such places in the series where deviation from the periodic law of the chemical properties of the elements are observed.

The major part of this paper concerned determination of the electron configurations of atoms other than hydrogen. Bohr was successful in calculating the approximate electron configurations for the six lightest elements, but his other determinations proved to be in error.

Although the Bohr model of the atom has undergone extensive modification, it nevertheless forms the basis for the model of the atom accepted today. The Bohr model led directly to the determination of electron configurations for the atoms of all the elements. This made it possible to explain the chemical properties of the elements and explain their systematic arrangement in the periodic table of Mendeleev. Bohr's explanation of the interaction of energy with matter as manifested in atomic spectra, removed the last major stumbling block to the full acceptance of the nuclear model of the atom.

Bohr's radical assumptions caused some concern among physicists. However, Rutherford's statement the following year (40) demonstrates that the importance of Bohr's work was recognized immediately.

> Bohr has drawn attention to the difficulties of constructing atoms on the "nucleus" theory, and has shown that the stable positions of the external electrons cannot be deduced from the classical mechanics. By the introduction of a conception connected with Planck's quantum, he has shown that on certain assumptions it is possible to construct simple atoms and molecules out of positive and negative nuclei, *e.g.,* the hydrogen atom and molecule and the helium atom, which behave in many respects like the actual atoms or molecules. While there may be much difference of opinion as to the validity and of the underlying physical meaning of the assumptions made by Bohr, there can be no doubt that the theories of Bohr are of great interest and importance to all physicists as the first definite attempt to construct simple atoms and molecules and to explain their spectra.

Chapter 5

Isotopes

The somewhat distinct lines of investigation outlined in chapters 2, 3 and 4 dovetailed together in 1913 to produce another startling discovery concerning the nature of the atom. Dalton's fundamental assumption that all the atoms of an element were identical in weight was disproved by this discovery.

In Chapter 2 experiments using the cathode ray tube were discussed at length. It was mentioned that Thomson and others had examined the positive rays developed in a modified cathode ray tube, and found that the charge to mass ratio of the particles indicated that the rays were composed of atoms with one or more electrons missing. Later experiments with improved positive ray equipment provided Thomson and his colleagues with startling new information about the structure of the atom.

In Chapter 3 we saw that Rutherford and Soddy established their theory of radioactivity and indicated that uranium and thorium were disintegrating into other substances, implying a *series* of radioactive transmutations for both elements. During the first 12 years of the 20th century, a tremendous effort was put forth by chemists and physicists in the United States and Europe to locate and isolate each successive member of all the radioactive series (a third series was discovered in the process). This radiochemistry history is a fascinating study in itself, but falls outside the scope of this book. By 1913 however, about 30 different radioactive substances had been identified. Many of these radioactive substances were chemically very similar or possibly identical, as evidenced by the fact that chemists could not separate them.

The evidence presented in Chapter 4 indicated that the atom possessed a nucleus which contained all the positive charge of the atom. It had also been conclusively demonstrated that the charge on the nucleus determined the chemical properties of an element, and ultimately, the number of possible elements.

Combining the evidence from Chapters 3 and 4 it appears that there is no room on the periodic table for the 30 radioactive substances which had been discovered! By the end of 1913, the solution to this major problem had been pretty well worked out.

Very early in 1913, J.J. Thomson announced some very interesting results from his positive ray studies. He had directed a beam of positive rays (a stream of charged atoms) through both electric and magnetic fields arranged so that all the particles of the same mass would fall in a parabolic curve on a photographic plate. Particles of different mass would show up as separate curves on the photographic plate (60).

> The method to which I shall refer this evening is the one I described in a lecture I gave here two years ago. The nature of the method may be understood from the diagram given in Figure 15. A is a vessel containing the gases at a very low pressure: an electric discharge is sent through these gases, passing from the anode to the kathode C. The positively electrified particles move with great velocity towards the kathode; some of them pass through a small hole in the centre, and emerge on the other side as a fine pencil of positively electrified particles. This pencil is acted on by electric forces when it

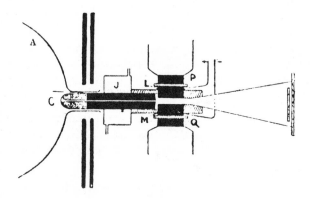

Figure 15.

> passes between the plates L and M, which are connected with the terminals of a battery of storage cells, and by a magnetic force when it passes between P and Q, which are the poles of an electromagnet.

In the pencil before it passed under the influence of these forces there might be many kinds of atoms or molecules, some heavy, others light, some moving quickly, others comparatively slowly, but these would all be mixed up together. When they are acted on by the electric and magnetic forces, however, they get sorted out, and instead of travelling along the same path they branch off into different directions. No two particles will travel along the same path unless they have the same mass as well as the same velocity; so that if we know the path of the particle we can determine both its mass and it velocity. In chemical analyses we are concerned more with the mass than with the velocity, and we naturally ask what is the connection between the paths of particles which have the same mass but move with different velocities. The answer is that all such paths lie on the surface of a cone, and that each kind of particle has its own cone; there is one cone for hydrogen, another for oxygen, and so on. Thus one cone is sacred to hydrogen, and if it exists there must be hydrogen in the vessel; so that if we can detect the different cones produced from the original pencil, we know at once the gases that are in the tube. Now, there are several ways of identifying these cones, but I shall only refer to the one I have used in the experiments I wish to bring before you this evening. These moving electrified particles, when they strike against a photographic plate, make an impression on the plate, and a record of the place where they struck the plate can be obtained. Thus, when a plate is placed in the way of the particles streaming along these cones, the sections of these cones by the plate (parabolas) are recorded on the photograph, hence we can identify these cones by the parabolic curves recorded on the photograph, and these parabolas will tell us what gases are in the vessel. . . .

Thomson then presented his discovery of a parabolic curve representing particles of atomic weight 22. Neon has an atomic weight of about 20 and Thomson suggests that the new substance might be a compound of neon and two hydrogen atoms. He doesn't suggest that the atoms of atomic weight 22 might actually be atoms of neon which were heavier than the more common neon atoms. That suggestion was to come later.

I now turn to the photograph of the lighter constituents shown in Figure 16; here we find the lines of helium, of neon (very strong), of argon, and, in addition, there is a line corresponding to an atomic weight 22, which cannot be identified with the line due to any known gas. I thought at first that this line, since its atomic weight is one-half that of CO_2, must be due to a carbonic acid molecule with a double charge of electricity, and on some of the

plates a faint line at 44 could be detected. On passing the gas slowly
through tubes immersed in liquid air the line at 44 completely disappeared, while the brightness of the one at 22 was not affected.

The origin of this line presents many points of interest; there
are no known gaseous compounds of any of the recognised elements
which have this molecular weight. Again, if we accept Mendeleef's
periodic law, there is no room for a new element with this atomic

Figure 16.

weight. The fact that this line is bright in the sample when the neon
line is extraordinarily bright, and invisible in the other when the
neon is comparatively feeble, suggests that it may possibly be a
compound of neon and hydrogen NeH_2, though no direct evidence
of the combination of these inert gases has hitherto been found.

A few months later he presented his somewhat clarified ideas (61).
He still does not suggest however, that the atoms of atomic weight
22 may be atoms of neon with a higher atomic weight.

"Neon," it is shown, is not a simple gas, but a mixture of two
gases, containing a large quantity of a gas of atomic weight about 20,
and a much smaller quantity of one with an atomic weight about 22.
The "22" gas was first observed in samples of residues of liquid air

supplied by Sir James Dewar, and has since been found in every specimen of neon examined, including a specimen supplied by M. Claud, of Paris, and a very carefully purified sample of neon prepared by Mr. Watson.

At about the same time, many of the scientists who were isolating radioactive substances began to conclude that atoms with the same nuclear charge, *i.e.,* atoms of the same element, could have different atomic weights.

Bohr, in his second 1913 paper (9) states this position very clearly.

> As is well known, several of these substances have very similar chemical properties and have hitherto resisted every attempt to separate them by chemical means. There is also some evidence that the substances in question show the same line-spectrum. It has been suggested by several writers that the substances are different only in radio-active properties and atomic weight but identical in all other physical and chemical respects. According to the theory, this would mean that the charge on the nucleus, as well as the configuration of the surrounding electrons, was identical in some of the elements, the only difference being the mass and the internal constitution of the nucleus.
>
> ... this assumption is already strongly suggested by the fact that the number of radioactive substances is greater than the number of places at our disposal in the periodic system. If, however, the assumption is right, the fact that two apparently identical elements emit β-particles of different velocities, shows that the β-rays as well as the α-rays have their origin in the nucleus.
>
> This view of the origin of α- and β-particles explains very simply the way in which the change in the chemical properties of the radioactive substances is connected with the nature of the particles emitted. The results of experiments are expressed in the two rules:—
>
> (1) Whenever an α-particle is expelled the group in the periodic system to which the resultant product belongs is two units less than that to which the parent body belongs.
>
> (2) Whenever a β-particle is expelled the group of the resultant body is 1 unit greater than that of the parent.

In December of 1913 Soddy demonstrated the existence of atoms with identical chemical properties, but different atomic weights and suggested the word "isotope" to describe these atoms (55).

> That the intra-atomic charge of an element is determined by its place in the periodic table rather than by its atomic weight, as concluded by A. van der Broek (*Nature*, November 27, p. 372), is

strongly supported by the recent generalisation as to the radio-elements and the periodic law. The successive expulsion of one α and two β particles in three radio-active changes in any order brings the intra-atomic charge of the element back to its initial value, and the element back to its original place in the table, though its atomic mass is reduced by four units. . . .

I regard van der Broek's view, that the number representing the net positive charge of the nucleus is the number of the place which the element occupies in the periodic table when all the possible places from hydrogen to uranium are arranged in sequence, as practically proved so far as the relative value of the charge for the members of the end of the sequence, from thallium to uranium, is concerned. We are left uncertain as to the absolute value of the charge, because of the doubt regarding the exact number of rare-earth elements that exist. If we assume that all of these are known, the value for the positive charge of the nucleus of the uranium atom is about 90. Whereas if we make the more doubtful assumption that the periodic table runs regularly, as regards numbers of places, through the rare-earth group, and that between barium and radium, for example, two complete long periods exist, the number is 96. In either case it is appreciably less than 120, the number were the charge equal to one-half the atomic weight, as it would be if the nucleus were made out of α particles only. Six nuclear electrons are known to exist in the uranium atom, which expels in its changes six β rays. Were the nucleus made up of α particles there must be thirty or twenty-four respectively nuclear electrons, compared with ninety-six or 102 respectively in the ring. If, as has been suggested, hydrogen is a second component of atomic structure there must be more than this. But there can be no doubt that there must be some, and that the central charge of the atom on Rutherford's theory cannot be a pure positive charge, but must contain electrons, as van der Broek concludes.

So far as I personally am concerned, this has resulted in a great clarification of my ideas, and it may be helpful to others, though no doubt there is little originality in it. The same algebraic sum of the positive and negative charges in the nucleus, when the arithmetical sum is different, gives what I call "isotopes" or "isotopic elements," because they occupy the same place in the periodic table. They are chemically identical, and save only as regards the relatively few physical properties which depend upon atomic mass directly, physically identical also. Unit changes of this nuclear charge, so reckoned algebraically, give the successive places in the periodic table. For any one "place," or any one nuclear charge, more than one number of electrons in the outer-ring system may exist, and in such a case the element exhibits variable valency. But such changes of number, or of

valency, concern only the ring and its external environment. There is
no in- and out-going of electrons between ring and nucleus.

Rutherford, a few months later, acknowledged the possible existence
of isotopes in a discussion of nuclear charge (40).

> On this view, the nucleus charge is a fundamental constant of
> the atom, while the atomic mass of an atom may be a complicated
> function of the arrangement of the units which make up the nucleus.
>
> It should be borne in mind that there is no inherent impos-
> sibility on the nucleus theory that atoms may differ considerably in
> atomic weight and yet have the same nucleus charge.

Then, two months later, Rutherford and Andrade published strong
evidence in support of Soddy's isotope theory (47).

> It thus appears that the nucleus charge of radium B is the same
> as that of lead, for the atomic number of radium B, deduced by
> Moseley's formula from the γ-ray spectrum, is that to be expected
> for lead, and the strong lines of the γ-ray spectrum of radium B seem
> to be coincident with those of lead. . . .
>
> These results confirm in an unexpected way the correctness of
> this deduction of Soddy and Fajans, and also give a definite verifiça-
> tion of the hypothesis that two elements of different atomic weights
> may have identical spectra and identical chemical properties. A simi-
> lar result has been recorded by Sir J.J. Thomson and Aston in their
> work indicating that neon consists of a mixture of two gases of
> atomic weights about 20 and 22. The theory of the nucleus atom
> affords a simple explanation of such a result; for the chemical and
> physical properties are for the most part determined by the charge
> on the nucleus, and are practically independent of the mass of the
> nucleus. The properties of radioactivity and gravitation belong
> mainly to the nucleus. The fact that radium B is radioactive while
> lead is not, shows that the constitution of the nucleus is different in
> the two cases, and this is borne out by the known difference in
> atomic weights.

Thus, by 1914, the existence of isotopes among the radioactive elements had been pretty well confirmed, and was strongly suspected
for some of the lighter elements due to Thomson's work.

The first World War brought about an almost complete cessa-
tion of experimental work on the atom, since most of the physicists
were occupied by wartime duties.

As a result, it was not until 1920 that Frederick Aston published the next major verification of the isotope theory (1). Aston

had continued Thomson's work with positive rays and now supplied convincing evidence that many of the light elements were mixtures of atoms with the same atomic number but different atomic weights.

In the atomic theory put forward by John Dalton in 1801 the second postulate was: "Atoms of the same element are similar to one another and equal in weight." For more than a century this was regarded by chemists and physicists alike as an article of scientific faith. The only item among the immense quantities of knowledge acquired during that productive period which offered the faintest suggestion against its validity was the inexplicable mixture of order and disorder among the elementary atomic weights. The general state of opinion at the end of last century may be gathered from the two following quotations from Sir William Ramsay's address to the British Association at Toronto in 1897:—

There have been almost innumerable attempts to reduce the differences between atomic weights to regularity by contriving some formula which will express the numbers which represent the atomic weights with all their irregularities. Needless to say, such attempts have in no case been successful. Apparent success is always attained at the expense of accuracy, and the numbers reproduced are not those accepted as the true atomic weights. Such attempts, in my opinion, are futile. Still, the human mind does not rest contented in merely chronicling such an irregularity; it strives to understand why such an irregularity should exist. . . . The idea . . . has been advanced by Prof. Schutzenburger, and later by Mr. Crookes, that what we term the atomic weight of an element is a mean; that when we say the atomic weight of oxygen is 16, we merely state that the average atomic weight is 16; and it is not inconceivable that a certain number of molecules have a weight somewhat higher than 32, while a certain number have a lower weight.

This idea was placed on an altogether different footing some ten years later by the work of Sir Ernest Rutherford and his colleagues on radioactive transformations. The results of these led inevitably to the conclusion that there must exist elements which have chemical properties identical for all practical purposes, but the atoms of which have different weights. This conclusion has been recently confirmed in a most convincing manner by the production in quantity of specimens of lead from radioactive and other sources, which, though perfectly pure and chemically indistinguishable, give atomic weights differing by amounts quite outside the possible experimental error. Elements differing in mass but chemically identical and therefore occupying the same position in the periodic table have been called "isotopes" by Prof. Soddy.

At about the same period as the theory of isotopes was being developed by the radio chemists at the heavy end of the periodic table an extremely interesting discovery was made by Sir J.J. Thomson, which carried the attack into the region of the lighter and nonradioactive elements. This was that, when positive rays from gases containing the element neon were analysed by electric and magnetic fields, results were obtained which indicated atomic weights roughly 20 and 22 respectively, the accepted atomic weight being 20·2. This naturally led to the expectation that neon might be a mixture of isotopes, but the weight 22 might possibly be due to other causes, and the method of analysis did not give sufficient accuracy to distinguish between 20·0 and 20·2 with certainty. Attempts were made to effect partial separation first by fractionation over charcoal cooled in liquid air, the results of which were absolutely negative, and then by diffusion, which in 1913 gave positive results, an apparent change in density of 0·7 per cent. between the lightest and heaviest fractions being attained after many thousands of operations. When the war interrupted the research, it might be said that several independent lines of reasoning pointed to the idea that neon was a mixture of isotopes, but that none of them could be said to carry the conviction necessary in such an important development.

By the time work was started again the isotope theory had been generally accepted so far as the radioactive elements were concerned, and a good deal of theoretical speculation had been made as to its applicability to the elements generally. As separation by diffusion is at the best extremely slow and laborious, attention was again turned to positive rays in the hope of increasing the accuracy of measurements to the required degree. This was done by means of the arrangement illustrated in Figure 17. Positive rays are sorted into

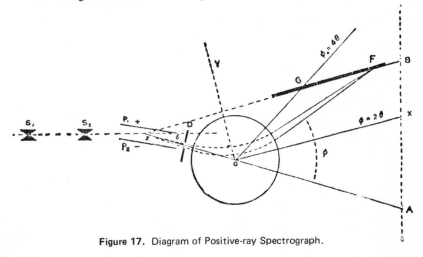

Figure 17. Diagram of Positive-ray Spectrograph.

an extremely thin ribbon by means of parallel slits S_1 S_2, and are then spread into an electric spectrum by means of the charged plates P_1 P_2. A portion of this spectrum deflected through an angle θ is selected by the diaphragm D and passed between the circular poles of a powerful electromagnet O the field of which is such as to bend the rays back again through an angle ϕ more than twice as great as θ. The result of this is that rays having a constant mass (or more correctly constant m/e) will converge to a focus F, and that if a photographic plate is placed at GF as indicated, a *spectrum dependent on mass alone* will be obtained. On account of its analogy to optical apparatus, the instrument has been called a positive-ray spectrograph and the spectrum produced a mass-spectrum. . . .

In spectrum i. the presence of neon is indicated by the lines 20 and 22 situated between these groups. Comparative measurements show that these lines are $20 \cdot 00$, $22 \cdot 00$, with an accuracy of one-tenth per cent, which removes the last doubt as to the isotopic nature of neon.

The next element investigated was chlorine; this is characterised by four strong lines 35, 36, 37, 38, and fainter ones at 39, 40; there is no trace of a line at $35 \cdot 46$, the accepted atomic weight. From reasoning which cannot be given here in detail it seems certain that chlorine is a complex element and consists of isotopes of atomic weights 35 and 37, with possibly another at 39. The lines at 36, 38 are due to the corresponding HCl's.

The work presented in this last paper together with follow-up work on other light elements dispelled any remaining doubts as to the existence of isotopes. Most of the lighter elements were, in fact, found to consist of two or more isotopes. A tentative explanation of isotopes was offered by Aston in a 1925 lecture (2). He assumed that a proton (hydrogen nucleus) and an electron can exist together in the nucleus. This neutral pair would add mass to the nucleus without affecting the net number of positive charges (atomic number).

In the fact that it is the charge and not the mass of the nucleus that controls the orbits of the planetary electrons we have the explanation of isotopes. Take a simple atom having 6 protons and 3 electrons in its nucleus, and 3 planetary electrons outside. Its nuclear charge, or atomic number, will be 3, and its atomic weight will be equal to the total number of protons it contains, that is 6. Now suppose into this atom is introduced a proton and an electron and further suppose that both enter the nucleus. The nucleus of the new atom will consist of 7 protons and 4 electrons. The net charge 3 will not be altered, and as this is the sole controlling force operating on the planetary electrons these will be unaware of any change at all and will behave exactly the same as before. The chemical and spectroscopic properties of the old and the new atoms will be identical, but whereas the atomic weight of the first is 6, that of the second is 7. These atoms will be isotopes,. . .

We shall see in the next chapter that the discovery of the neutron in 1932 produced the final explanation of isotopes.

Chapter 6

Transmutation: The Discovery and Isolation of the Proton and Neutron

In the preceding chapters we traced the development of a coherent nuclear model of the atom which was capable of explaining a wealth of experimental results. In this chapter we will follow the experimental scientists in their attempt to determine the structure of the nucleus itself. First, we will follow the amazing experimental procedures of Rutherford which produced the first experimentally induced transmutation of elements and led to the discovery and isolation of the proton. In the second part of the chapter we will review the history of scientific thought concerning the structure of the nucleus which culminated in 1932 with the discovery and isolation of the neutron.

The experimental approach which led Rutherford to yet another phenomenal scientific breakthrough involved the bombardment of gases with α particles. Marsden began the investigations with a few simple experiments (22).

> On the nucleus theory of an atom, Sir Ernest Rutherford and C.G. Darwin have recently discussed the relative motions of an α particle and the nucleus of an atom in an intimate collision. In the case of the encounter of an α particle ... with the nucleus of a hydrogen atom, ... it can be deduced that in an end-on collision, ... the "H" particle will have about four times the range of the α particle producing it. Consequently, in the passage of α particles through hydrogen the "H" particles may be looked for well beyond the range of the ordinary α particles, and in the experiments to be described evidence of their existence has been found.

Rutherford elaborated on these results (40).

> *Scattering in Hydrogen.*
>
> Special interest attaches to the effects to be expected when α particles pass through light gases like hydrogen and helium....
>
> It is only necessary here to refer to the fact that on the nucleus theory a small number of hydrogen atoms should acquire, as

the result of close encounters with α particles, velocities about 1·6 times that of the velocity of the α particle itself. On account of the fact that the hydrogen atom carries one positive charge while the α particle carries two, it can be calculated that some of the hydrogen atoms should have a range in hydrogen of nearly four times that of the α particle which sets them in motion.

Mr. Marsden has kindly made experiments for me to test whether the presence of such hydrogen atoms can be detected. A detailed account of his experiments will appear later, but it suffices to mention here that undoubted evidence has been obtained by him that some of the hydrogen atoms are set in such swift motion that they are able to produce a visible scintillation on a zinc sulphide screen and are able to travel through hydrogen a distance three or four times greater than the colliding α particle. . . .

, the great majority of the scintillations disappeared at about 20 cm. from the source, which corresponds to the range of the α particle in hydrogen. A small number of scintillations, however, persisted in hydrogen up to a distance of about 90 cm. The scintillations were of less intensity than those due to the ordinary α particle. . . .

There appears to be no doubt that the scintillations observed beyond 20 cm. are due to charged hydrogen atoms which are set in swift motion by a close encounter with an α particle.

Further experiments began to produce some very strange results. In the experiments presented next, it appeared that hydrogen nuclei were being emitted by radioactive source materials (this later proved to be incorrect). Nothing of this sort had been previously recorded, so naturally, Rutherford became interested. Marsden and Lantsberry report their experimental results (23).

In a previous paper by one of us, evidence was given that in the passage of α particles through hydrogen, certain intimate collisions take place with the nuclei of the hydrogen atoms. These encounters result in the nuclei attaining velocities greater than those of the α particles impinging on them, giving them a correspondingly greater penetrating power or range. . . .

It seemed desirable, however, to make quantitative measurements of the number and distribution of the H particles when a parallel beam of α particles of definite velocity and intensity falls on a known thickness of hydrogen gas. . . .

A pencil of α particles from an α-ray tube containing radium emanation fell on a thin sheet (10μ) of wax, and the H particles projected from the hydrogen in the wax were counted in different directions by means of a zinc-sulphide screen. . . .

The effects observed were several times greater than anticipated by formula. It was found, however, that the source itself was
emitting long-range particles capable of scintillating, and that these
were scattered in the wax, thus causing the disturbance. Consequently an investigation of these long-range particles had to be
undertaken.

Long-Range Particles emitted from the source of Radium Emanation.

The α-ray tube was placed in air at a distance of 8 cm. from a
zinc-sulphide screen, the path between the two being in a transverse
magnetic field to get rid of disturbing luminosity due to β particles.
Although the α particles did not penetrate more than 5·8 cm. from
the α-ray tube, yet scintillations were observed on the screen. These
scintillations were similar in appearance to those produced by H
particles. On moving the zinc-sulphide screen further from the
source, the numbers fell off in the same way as H particles might be
expected to be absorbed. Further, on interposing sheets of aluminium between the source and screen, the curve of absorption was the
same, within the experimental error, as the curve for H particles
given in the previous paper. These results show that H particles are
given off from the source, and that, unlike α particles, their velocities are not uniform, but are distributed in the same manner as
those of H particles produced during the ordinary transmission of α
particles through hydrogen. . . .

Thus there seems a strong suspicion that H particles are
emitted from the radioactive atoms themselves, though not with
uniform velocity.

The first World War disrupted most experimental work, but
Rutherford was very intrigued by Marsden's results and when time
permitted, continued and expanded Marsden's experiments. In an
amazing series of four papers, published together in a single issue of
the Philosophical Magazine, we can trace the development of Rutherford's ideas as he obtained his surprising experimental results. In the
fourth of these papers, he announced the first artificial transmutation of elements. The alchemists' dream of changing one element
into another really came true!

The first paper (41) presented a description of the apparatus
used throughout his experiments, but contained no very startling
results.

On the nucleus theory of atomic structure, it is to be anticipated that the nuclei of light atoms should be set in swift motion by
intimate collisions with α particles. From consideration of impact, it

can be simply shown that as a result of a head-on collision, an atom of hydrogen should acquire a velocity 1·6 times that of the α particle before impact, and should possess ·64 of the energy of the incident α particle. Such high speed "H" atoms should be readily detected by the scintillation method. This was shown to be the case by Marsden, who found that the passage of α particles through hydrogen gave rise to numerous faint scintillations on a zinc sulphide screen placed far beyond the range of the α particles. The maximum range of the H particles, set in motion by the α particles from radium C, was over 100 cm. in hydrogen or about four times the range of the colliding α particles in that gas. . . .

In a second paper, Marsden showed that the α-ray tube itself gave rise to a number of scintillations like those from hydrogen. . . .

The number of H scintillations observed in all cases appeared to be too large to be accounted for by the possible presence of hydrogen in the material, and Marsden concluded that there was strong evidence that hydrogen arose from the radioactive matter itself. . . .

We have seen that Marsden in his second paper had some indications that the radioactive matter itself gave rise to swift H atoms. This, if correct, was a very important result, for previously the presence of no light element except helium had been observed in radioactive transformations.

It was thought desirable to continue these experiments in more detail, and during the past four years I have made a number of experiments on this point and on other interesting problems that have arisen during the progress of the work. . . .

Experimental arrangement.

For experiments with hydrogen and other gases, the active disk D (Figure 18) was mounted at a convenient height parallel to the screen on a metal bar B which slid into a rectangular brass box A, 18 cm. long, 6 cm. deep, and 2 cm. wide, with metal flanges at both ends fitting between the rectangular poles of a large electromagnet. One end was closed by a ground glass plate C, and the other by a waxed brass plate E, in the centre of which was cut a rectangular opening 1 cm. long and 3 mm. wide. This opening was covered by a thin plate of metals of silver, aluminium or iron, whose stopping power for α particles lay between 4 and 6 cm. of air. The zinc sulphide screen F was fixed opposite the opening and distant 1 or 2 mm. from the metal covering. By means of two stopcocks, the vessel was filled with the gas to be examined either by exhaustion or displacement. . . .

Since the number of H atoms observed under ordinary conditions is less than one in a hundred thousand of the number of α particles, H atoms, projected in the direction of the α particles, can

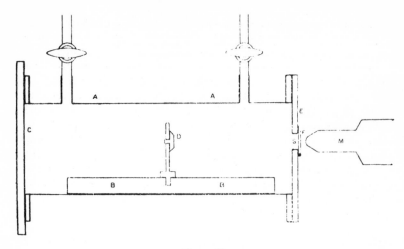

Figure 18.

only be detected when the α rays are stopped by the absorbing screens. . . .

It is clear from the results given in this paper that a close collision between an α particle and a hydrogen nucleus is an exceedingly rare occurrence. Only 1 in 100000 of the α particles passing through 1 cm. of hydrogen at normal pressure and temperature gives rise to a high-speed H atom, while in the same distance each α particle on an average passes through the sphere of action of about 10000 hydrogen molecules. Thus for every 10^9 collisions with the molecules, in only one case does the α particle pass close enough to the nucleus to give rise to a swift H atom. . . .

It is clear that for such close collisions, each hydrogen atom in any complex molecule acts as an independent unit, so that swift H atoms should be liberated by α particles from every substance containing free or combined hydrogen. This is fully borne out by experiment.

In his summary, Rutherford comments on Marsden's previous results.

As observed by Marsden, hydrogen atoms are emitted by the radioactive source. The number observed is small, and it is difficult to decide whether these H atoms arise from the radioactive transformation or from occluded hydrogen in the source.

In the second paper (42) Rutherford, for the first time, suggests that particle bombardment might cause disintegration of atomic nuclei.

In the first paper giving an account of the number of H atoms produced by α particles and their absorption by matter, it has been implicitly assumed that the long-range scintillations observed in hydrogen are due to swift hydrogen atoms set in motion by close collisions with α particles. This is supported by the observations that the range of the atoms is in good accord with the value calculated by Darwin from Bohr's theory of absorption of charged particles.

Taking into account, however, the intense forces developed in such collisions and the possibility of the disruption of the structure of the nuclei involved in the collisions, it was thought desirable to determine experimentally the mass and velocity of these flying atoms, and to compare the values with those deduced from the collision theory. Such a determination was rendered the more necessary by certain apparent anomalies observed in connexion with the brightness and distribution of H atoms, an account of which will be given later in this paper.

To determine the mass and velocity of the H atom, it was necessary to measure the deflexions of a stream of H atoms both in a magnetic and in an electric field. The experiments were somewhat tedious and difficult on account of the small number of H scintillations present under the experimental restrictions. . . .

. . . consequently, $e/m = 10^4$ e.m. units.

The value of e/m for the hydrogen atom in the electrolysis of water is 9570. The agreement is sufficiently close to show that the long-range scintillations produced by α particles in hydrogen are due to hydrogen atoms carrying a unit positive charge. The agreement between the calculated and observed velocities shows that, within the margin of experimental error, the conservation of momentum and energy hold for close collisions between the atomic nuclei and that there is no sensible loss of energy due to radiation.

In the third paper (43) Rutherford turned his attention to the collision of α particles with nitrogen and oxygen atoms. He reports some unexpected experimental results, but gives no real hint of the amazing results presented in the fourth paper.

In a previous paper we have calculated the number and distribution of H atoms produced by α particles on the assumption that the nuclei may be regarded as point centres of force repelling according to the law of the inverse square. When these calculations are applied to the collision of α particles with nitrogen or oxygen nuclei, the distribution with velocity of the N and O atoms is very similar to that for H atoms. We should consequently expect on the simple theory that the number of N and O atoms should fall off very rapidly between 7 and 9 cm., and that the number of short-range atoms should greatly preponderate. Quite the contrary is observed in the experiments . . . , where it is seen that the number of scintillations fall off quite gradually with range. . . .

We may consequently conclude that about the same number of swift atoms are produced per centimetre of path by the passage of α particles through air, oxygen, and hydrogen. As in the case of hydrogen, it can be shown that all α particles, shot within a perpendicular distance $p = 2 \cdot 4 \times 10^{-13}$ cm. of the atomic nucleus, give rise to swift atoms of nitrogen and oxygen.

The fourth paper in this series is a real classic. Rutherford observed a very unexpected experimental result: the bombardment of dry air by α particles produced long-range particles. His tenacious follow up of this observation, leading him finally to conclude that a transmutation of nitrogen had occurred, is a brilliant example of scientific technique carried out on the highest level (44).

It has been shown in paper I. that a metal source, coated with a deposit of radium C, always gives rise to a number of scintillations on a zinc sulphide screen far beyond the range of the α particles. The swift atoms causing these scintillations carry a positive charge and are deflected by a magnetic field, and have about the same range and energy as the swift H atoms produced by the passage of α particles through hydrogen. These "natural" scintillations are believed to be due mainly to swift H atoms from the radioactive source, but it is difficult to decide whether they are expelled from the radioactive source itself or are due to the action of α particles on occluded hydrogen.

The apparatus employed to study these "natural" scintillations is the same as that described in paper I. The intense source of radium C was placed inside a metal box about 3 cm. from the end, and an opening in the end of the box was covered with a silver plate of stopping power equal to about 6 cm. of air. The zinc sulphide screen was mounted outside, about 1 mm. distant from the silver plate, to admit of the introduction of absorbing foils between them. The whole apparatus was placed in a strong magnetic field to deflect the β rays. The variation in the number of these "natural" scintillations with absorption in terms of cms. of air is shown in Figure 1, curve A. In this case, the air in the box was exhausted and absorbing foils of aluminium were used. When dried oxygen or carbon dioxide was admitted into the vessel, the number of scintillations diminished to about the amount to be expected from the stopping power of the column of gas.

Now came the surprise!.

A surprising effect was noticed, however, when dried air was introduced. Instead of diminishing, the number of scintillations was increased, and for an absorption corresponding to about 19 cm. of air the number was about twice that observed when the air was exhausted. It was clear from this experiment that the α particles in their passage through air gave rise to long-range scintillations which appeared to the eye to be about equal in brightness to H scintillations. A systematic series of observations was undertaken to account for the origin of these scintillations. In the first place we have seen that the passage of α particles through nitrogen and oxygen gives rise to numerous bright scintillations which have a range of about 9 cm. in air. These scintillations have about the range to be expected if they are due to swift N or O atoms, carrying unit charge, produced by collision with α particles. All experiments have consequently been made with an absorption greater than 9 cm. of air, so that these atoms are completely stopped before reaching the zinc sulphide screen.

It was found that these long-range scintillations could not be due to the presence of water vapour in the air; for the number was only slightly reduced by thoroughly drying the air. . . .

Even when oxygen and carbon dioxide saturated with water vapour at 20°C. were introduced in place of dry air, the number of scintillations was much less than with dry air.

It is well known that the amount of hydrogen or gases containing hydrogen is normally very small in atmospheric air. No difference was observed whether the air was taken directly from the room or from outside the laboratory or was stored for some days over water.

There was the possibility that the effect in air might be due to liberation of H atoms from the dust nuclei in the air. No appreciable difference, however, was observed when the dried air was filtered through long plugs of cotton-wool, or by storage over water for some days to remove dust nuclei.

Since the anomalous effect was observed in air, but not in oxygen, or carbon dioxide, it must be due either to nitrogen or to one of the other gases present in atmospheric air. The latter possibility was excluded by comparing the effects produced in air and in chemically prepared nitrogen. The nitrogen was obtained by the well-known method of adding ammonium chloride to sodium nitrite, and stored over water. It was carefully dried before admission to the apparatus. With pure nitrogen, the number of long-range scintillations under similar conditions was greater than in air. As a result of careful experiments, the ratio was found to be 1·25, the value to be expected if the scintillations are due to nitrogen.

The results so far obtained show that the long-range scintillations obtained from air must be ascribed to nitrogen, but it is important, in addition, to show that they are due to collision of α particles with atoms of nitrogen through the volume of the gas. In the first place, it was found that the number of the scintillations varied with the pressure of the air in the way to be expected if they resulted from collision of α particles along the column of gas. In addition, when an absorbing screen of gold or aluminium was placed close to the source, the range of the scintillations was found to be reduced by the amount to be expected if the range of the expelled atom was proportional to the range of the colliding α particles. These results show that the scintillations arise from the volume of the gas and are not due to some surface effect in the radioactive source. . . .

Both as regards range and brightness of scintillations, the long-range atoms from nitrogen closely resemble H atoms, and in all probability are hydrogen atoms. In order, however, to settle this important point definitely, it is necessary to determine the deflexion of these atoms in a magnetic field. Some preliminary experiments have been made by a method similar to that employed in measuring the velocity of the H atom . . .

As far as the experiment goes, this is an indication that the scintillations are due to H atoms; but the actual number of scintillations and the amount of reduction was too small to place much reliance on the result. In order to settle this question definitely, it will probably prove necessary to employ a solid nitrogen compound, free from hydrogen, as a source, and to use much stronger sources of α rays. . . .

Discussion of results.

From the results so far obtained it is difficult to avoid the conclusion that the long-range atoms arising from collision of α particles with nitrogen are not nitrogen atoms but probably atoms of hydrogen, or atoms of mass 2. If this be the case, we must conclude that the nitrogen atom is disintegrated under the intense forces developed in a close collision with a swift α particle, and that the hydrogen atom which is liberated formed a constituent part of the nitrogen nucleus. We have drawn attention in paper III. to the rather surprising observation that the range of the nitrogen atoms in air is about the same as the oxygen atoms, although we should expect a difference of about 19 per cent. If in collisions which give rise to swift nitrogen atoms, the hydrogen is at the same time disrupted, such a difference might be accounted for, for the energy is then shared between two systems. . . .

We should anticipate from radioactive data that the nitrogen nucleus consists of three helium nuclei each of atomic mass 4 and either two hydrogen nuclei or one of mass 2. If the H nuclei were outriders of the main system of mass 12, the number of close collisions with the bound H nuclei would be less than if the latter were free, for the α particle in a collision comes under the combined field of the H nucleus and of the central mass. Under such conditions, it is to be expected that the α particle would only occasionally approach close enough to the H nucleus to give it the maximum velocity, although in many cases it may give it sufficient energy to break its bond with the central mass. Such a point of view would explain why the number of swift H atoms from nitrogen is less than the corresponding number in free hydrogen and less also than the number of swift nitrogen atoms. The general results indicate that the H nuclei, which are released, are distant about twice the diameter of the electron (7×10^{-13} cm.) from the centre of the main atom. . . .

the close collision of such as α particle with a light atom seems to be the most likely agency to promote the disruption of the latter; for the forces on the nuclei arising from such collisions appear to be greater than can be produced by any other agency at present available. Considering the enormous intensity of the forces brought into play, it is not so much a matter of surprise that the nitrogen atom should suffer disintegration as that the α particle itself escapes disruption into its constituents. The results as a whole suggest that, if α particles—or similar projectiles—of still greater energy were available for experiment, we might expect to break down the nucleus structure of many of the lighter atoms.

Hence, Rutherford concluded that hydrogen atoms or atoms of mass 2 are released from the nitrogen nucleus during its collision with an α particle. The experimental results do not, however, suggest what happens to the α particle involved in the collision. Does it bounce away after the collision or does it get captured and become part of the nitrogen nucleus? This question was not answered until five years later. Nevertheless, the nucleus of a small atom, nitrogen, had been disrupted, and scientists had a new tool for examining the structure of the atom.

A hydrogen atom nucleus was apparently released from the nucleus of the nitrogen atom during collision with an α particle. Also, no positively charged particle had been isolated with less charge or less mass than the hydrogen nucleus. It is not surprising then, that scientists soon suspected that the hydrogen nucleus is a fundamental particle which provides a vast majority of the mass and positive charge of all atomic nuclei.

Rutherford, in 1922 (46) defended this position and suggested the name "proton" for this particle.

We shall see that only a small fraction of the mass of an atom can be ascribed to the negative electrons contained in it, but the main part is to be ascribed to the positively charged units of its structure. One of the main difficulties in our attack on the question of atomic constitution has lain in the uncertainty of the nature of positive electricity. Without entering upon the changes in point of view on this important question, it may suffice to say that the evidence as a whole supports the idea that the nucleus of the hydrogen atom, *i.e.* a positively charged atom of hydrogen, is the positive electron. No evidence has been obtained of the existence of a positively charged unit of mass less than that of the hydrogen nucleus, either in vacuum tubes or in the transformation of the radioactive atoms, where the processes occurring are very fundamental in character.

It might *a priori* have been anticipated that the positive electron should be the counterpart of the negative electron and have the same small mass. There is, however, not the slightest evidence of the existence of such a counterpart. On the views oulined, the positive and negative electrons both consist of the fundamental unit of charge, but the mass of the positive is about 1800 times that of the negative. This difference in the mass of the two electrons seems a fundamental fact of nature and, indeed, is essential for the existence of atoms as we know them. The unsymmetrical distribution of positive and negative electricity that is characteristic for all atoms is a consequence of this wide difference in the mass of the ultimate electrons which compose their structure. No explanation can be offered at the moment why such a difference should exist between positive and negative electricity.

Since it may be argued that a positive unit of electricity associated with a much smaller mass than the hydrogen nucleus may yet be discovered, it may be desirable not to prejudge the question by calling the hydrogen nucleus the positive electron. For this reason, and also for brevity, it has been proposed that the name "proton" should be given to the unit of positive electricity associated in the free state with a mass about that of the hydrogen nucleus, namely, about $1 \cdot 007$ in terms of $O = 16$. A name for this unit will be found very convenient in discussing the inner structure of atoms. In the following, the term "electron" will be applied only to the well-known negative unit of electricity of small mass.

By 1924 further experiments by Rutherford and Chadwick showed that in collisions with alpha particles, atoms of light elements other than nitrogen, release protons. They were unsuccessful, however, in their attempts to disrupt the nuclei of heavy atoms (48).

In previous papers we have shown that hydrogen nuclei are ejected from the elements boron, nitrogen, fluorine, sodium, aluminium, and phosphorus by bombardment with α particles. In these experiments the material subjected to the bombardment was placed immediately in front of the source of α particles and observations of the ejected particles were made on a zinc sulphide screen placed in a direct line a few centimetres away. . . .

Working in this way we have found that in addition to the elements boron, nitrogen, fluorine, sodium, aluminium, and phosphorus, which give H particles of maximum range in the forward direction between 40 and 90 cm., the following give particles of range above 7 cm.: neon, magnesium, silicon, sulphur, chlorine, argon, and potassium. . . .

We hope later to make a systematic examination of the elements with an improved counting microscope in order to settle definitely whether any evidence of disintegration can be obtained.

The final dramatic step in the isolation of the proton came in 1925 as the result of the marvelous photographic techniques of P. M. S. Blackett (7). He presented photographic evidence of the existence of the proton and resolved the question of what happened to the α particle after the collision with a nitrogen nucleus.

The original experiments of Rutherford and later those of Rutherford and Chadwick have shown that fast alpha-particles are able by close collisions to eject protons from the nuclei of many light elements. In particular the protons from boron, nitrogen, fluorine, sodium, aluminium and phosphorus have great ranges; and are emitted in all directions relative to the velocity of the bombarding alpha-particles. The scintillation method used in these experiments can give no direct information about the motion after the collision of the residual nucleus and of the alpha-particle itself. The proton alone has sufficient range to make detection by the scintillation method possible. The Wilson Condensation Method provides the obvious and perhaps the only certain way of observing the motion of these two particles. Of the "active" elements mentioned, nitrogen can at once be selected as the most suitable for a first investigation. . . .

The total number* emitted in all directions by a million 8·6 cm. alpha-particles can be estimated. . . .

In order to photograph a large number of tracks, a modified and automatic form of Wilson's** apparatus was constructed, which made one expansion and took one photograph every ten or fifteen seconds. . . .

The camera, designed originally by Shimizu, takes two photographs at right angles on standard cinematograph film.

*(of protons)
**(cloud-chamber)

About 23,000 photographs have been taken of the tracks of alpha-particles in nitrogen. . . .

The average number of tracks on each photograph was 18; the tracks of about 270,000 alpha-particles of 8·6 cm. range and 145,000 of 5·0 cm. range have therefore been photographed. . . .

Amongst these tracks a large number of forks were found corresponding to the elastic collisions make by alpha-particles with nitrogen (and oxygen) atoms. Reproductions of a few such tracks are given . . .

But amongst these normal forks due to elastic collisions, eight have been found of a strikingly different type. Six of them are reproduced. . . . These eight tracks undoubtedly represent the ejection of a proton from a nitrogen nucleus. It was to be expected that a photograph of such an event would show an alpha-ray track branching into three. The ejected proton, the residual nucleus from which it has been ejected, and the alpha-particle itself, might each have been expected to produce a track. These eight forks however branch only into two. The path of the first of the three bodies, the ejected proton, is obvious in each photograph. It consists of a fine straight track, along which the ionisation is clearly less than along an alpha-ray track, and must therefore be due to a particle of small charge and great velocity. The second of the two arms of the fork is a short track similar in appearance to the track of the nitrogen nucleus in a normal fork. Of a third arm to correspond to the track of the alpha-particle itself after the collision there is no sign. On the generally accepted view, due to the work of Rutherford, the nucleus of an atom is so small, and thus the potential at its surface so large, that a positively charged particle that has once penetrated its structure (and almost certainly an alpha-particle that ejects a proton must do so) cannot escape without acquiring kinetic energy amply sufficient to produce a visible track. As no such track exists the alpha-particle cannot escape. In ejecting a proton from a nitrogen nucleus the alpha-particle is therefore itself bound to the nitrogen nucleus. The resulting new nucleus must have a mass 17, and, provided no electrons are gained or lost in the process, an atomic number 8. The possibility of such a capture has already been suggested by Rutherford, and Chadwick in a recent paper.

The argument so far has been based on the appearance of these anomalous tracks. The conclusions already drawn from their appearance are fully confirmed by measurement. . . .

In marked contrast to the normal forks, the angles between the components of each of these anomalous forks are not in general consistent with an *elastic* collision between an alpha-particle and a nucleus of any known or possible (*i.e.* integral) mass. . . .

This first photograph, 7 (taken from two different angles), shows the elastic collision of an α particle and a nitrogen atom. Notice how heavy the branches of the forks are. No nuclear reaction took place.

The beaded appearance of the proton tracks, evidence of the small ionisation along them, can be seen in the photographs. Since the ionisation due to any particle is proportional to the square of its charge and the reciprocal of its velocity, the ionisation density along these proton tracks should be about one-sixth of that along the alpha-ray tracks. . . .

In the second photograph, 1 (also taken from two different angles), a collision of an α particle and a nitrogen nucleus has produced a proton and an oxygen atom. An artificially induced transmutation has taken place.

Discussion of Results.

The study of the photographs has led to the conclusion that an alpha-particle that ejects a proton from a nitrogen nucleus is itself bound to that nucleus. This result is of such importance that it is useful to emphasise the evidence on which it is based.

The first step in the argument must show that the eight anom-

alous forks do actually represent the ejection of a proton from a
nitrogen nucleus. Their appearance makes this probable; the measurements of the forks, the frequency of their occurrence and the
absence of any other abnormal forks, make it certain.

The second step must show that if the alpha-particle is not
bound to the nitrogen nucleus after the collision, a third arm to the
forks would be found. . . . Rutherford was able to detect a minimum
range of protons ejected from aluminium and sulphur, corresponding
to a maximum potential of about 3×10^6 volts. Unless nitrogen is
markedly different from these elements, this then is the order of the
total kinetic energy of the alpha-particle and nucleus after separating, since it seems certain that to eject a proton an alpha-particle
must penetrate within the surface of maximum potential. Particles
of this total kinetic energy will certainly both produce visible tracks.
The fact that only one track is actually found makes it therefore
certain that the two particles are bound together. . . .

Of the nature of the integrated nucleus little can be said without further data. It must however have a mass 17, and provided no
other nuclear electrons are gained or lost in the process, an atomic
number 8. It ought therefore to be an isotope of oxygen.

Blackett's conclusions can be summed up in equation form:

$$ {}^{14}_{7}\text{N} \quad + \quad {}^{4}_{2}\text{He} \quad \longrightarrow \quad {}^{1}_{1}\text{H} \quad + \quad {}^{17}_{8}\text{O} $$

The superscript gives the atomic weight of the atom, while the sub-
script gives the atomic number. A conservation of both mass and
charge is assumed. The alchemist's dream of artificially induced
transmutation came true!

In the remainder of this chapter we will trace the development
of scientific thought on the structure of the nucleus. In the first
decades of the 19th century, an English physicist named Prout had
suggested that the atoms of elements heavier than hydrogen were
made up of specific numbers of hydrogen atoms bound together. He
thus viewed hydrogen as the fundamental building block of which all
the other elements were constructed. This theory later fell into dis-
repute when more accurate atomic weight determinations showed
that few elements had atomic weights which were exact multiples of
hydrogen's weight. In fact, some elements such as chlorine had
atomic weights which fell almost midway between multiples of
hydrogen's atomic weight. Enough elements were close to whole

number multiples of the atomic weight of hydrogen, however, to leave the question open. The burst of scientific activity we have witnessed at the beginning of the 20th century refocused interest on Prout's theory. As evidence accumulated it began to look as if he had been correct. In 1913 an avalanche of new evidence descended. The discovery that the nuclear charge increases in discrete jumps in successive elements in the periodic table and the fact that no particle had been found with less positive charge than the hydrogen atom, certainly supported Prout's theory. The recognition of isotopes the same year destroyed the last major objection to the theory. The fractional atomic weights, e.g., chlorine 35.45, could be explained as resulting from a mixture of isotopes, each having whole number multiples of the weight of the hydrogen atom. Finally, the proton, or hydrogen nucleus, was isolated in 1925 by Blackett. The biggest question remaining concerned the difference between the atomic number of an element and its atomic weight. What could account for this disparity? Van den Broek's comment below (65), shows that by 1913 the idea that the nucleus not only contained protons, but also electrons was already prevalent. The helium nucleus for example, would contain 4 protons and 2 electrons, giving a mass of 4, but a net positive charge of two.

> But even then the nucleus might contain electrons. If the*
> particle should, as probable, consist of 4(H+) and 2 electrons.

Rutherford commented on this idea the following year. The extreme stability of the α particle (helium nucleus) prompted him to suggest that it may be a structural unit in the nuclei of bigger atoms (40).

> The helium nucleus has a mass nearly four times that of hydrogen. If one supposes that the positive electron, *i.e.*, the hydrogen atom, is a unit of which all atoms are composed, it is to be anticipated that the helium atom contains four positive electrons and two negative.
> It is well known that a helium atom is expelled in many cases in the transformation of radioactive matter, but no evidence has so far been obtained of the expulsion of a hydrogen atom. . . .
> It thus follows that the helium nucleus is a very stable configuration which survives the intense disturbances resulting in its expulsion with high velocity from the radioactive atom, and is one

*(α)

of the units, of which possibly the great majority of the atoms are composed. . . .

An important question arises whether the atomic nuclei, which all carry a positive charge, contain negative electrons. This question has been discussed by Bohr, who concluded from the radioactive evidence that the high speed β particles have their origin in the nucleus. The general radioactive evidence certainly supports such a conclusion. It is well known that the radioactive transformations which are accompanied by the expulsion of high speed β particles are, like the α ray changes, unaffected by wide ranges of temperature or by physical and chemical conditions. On the nucleus theory, there can be no doubt that the α particle has its origin in the nucleus and gains a great part, if not all, of its energy of motion in escaping from the atom. It seems reasonable, therefore, to suppose that a β ray transformation also originates from the expulsion of a negative electron from the nucleus.

Rutherford in 1922, taking into consideration the information from his transmutation experiments and Aston's isotope work, again summed up current ideas on the structure of the nucleus (46).

From the radio-active evidence it seems clear that the nuclear structure contains both helium nuclei and electrons. In the uranium-radium series of transformations, eight helium nuclei are emitted and six electrons and it is natural to suppose that the helium nuclei and electrons that are ejected act as units of the nuclear structure. It is clear from these results that the nuclear charge of an element is the excess of the positive charges in the nucleus over the negative. It is a striking fact that no protons (H nuclei) appear to be emitted in any of the radio-active transformations, but only helium nuclei and electrons.

Some very definite and important information on the structure of nuclei has been obtained by Aston in his experiments to show the existence of isotopes in the ordinary stable elements by the well-known positive-ray method. He found that a number of the elements were simple and contained no isotopes. Examples of such "pure" elements are carbon, nitrogen, oxygen, and fluorine. It is significant that the atomic weights of these elements are nearly whole numbers in terms of $O = 16$; on the other hand elements such as neon, chlorine, krypton, and many others consisted of mixtures of two or more isotopes of different atomic masses. Aston found that within the limit of error—about 1 in 1000—the atomic weights of these isotopes were whole numbers on the oxygen scale. This is a very important result, and suggests that the nuclei of elements are built up by the addition of protons, of mass nearly one, in the nuclear combination.

> From the radio-active evidence, we know that the nuclei of heavy atoms are built up, in part at least of helium nuclei and electrons, while it also seems clear that the proton can be released from the nuclei of certain light atoms. It is, however, very natural to suppose that the helium nucleus which carries two positive charges is a secondary building unit, composed of a close combination of protons and electrons, namely, 4 protons and 2 electrons.
>
> From the point of view of simplicity, such a conception has much in its favour, although it should be mentioned that it seems at the moment impossible to prove its correctness.

The idea that the electrons exist both in the nucleus and in orbit around the nucleus was summed up by Aston in 1925 (2).

> Thus the atom of hydrogen consists of one proton and one electron; that of helium of four protons and four electrons; the lighter isotope of lithium six of each, the heavier isotope seven of each, and so on through the list of the elements.
>
> We will now consider the manner in which these unit charges of electricity, the heavy positive particles and the lighter negative ones, are grouped in the atom itself. Of the many theories of atomic structure put forward since the discovery of radio-activity one only has stood the test of time and is now generally accepted. This is the 'nucleus' theory which we owe to Sir Ernest Rutherford. Although this structure of the atom is incompatible with many of the laws of classical dynamics, not only is it generally accepted, but in consideration of the severity of the tests applied in modern research, it is hardly going too far to state that it must be substantially correct. On this view in the normal atom all the protons and about half the electrons are packed together into a massive central nucleus or sun around which circulate the remaining electrons.

This view of nuclear structure, envisioning combinations of protons and electrons to account for the neutral mass in the nucleus, was widely accepted. There were, however, some severe theoretical problems involved with this solution, and a few scientists had begun to speculate on the possible existence of a fundamental particle with about the same mass as the proton, but which had a neutral charge. Usually this speculation involved some combination of a proton and an electron. The most important speculations had been made by none other than Rutherford in a 1920 lecture (45). He was prophetic indeed, for not only did he correctly suggest the existence of a neutron and correctly assess its properties, but also suggested the existence of the heavy isotope of hydrogen which was discovered 11 years later.

... it seems very likely that one electron can also bind two H nuclei and possibly also one H nucleus. In the one case, this entails the possible existence of an atom of mass nearly 2 carrying one charge, which is to be regarded as an isotope of hydrogen. In the other case, it involves the idea of the possible existence of an atom of mass 1 which has zero nucleus charge. Such an atomic structure seems by no means impossible. On present views, the neutral hydrogen atom is regarded as a nucleus of unit charge with an electron attached at a distance, and the spectrum of hydrogen is ascribed to the movements of this distant electron. Under some conditions, however, it may be possible for an electron to combine much more closely with the H nucleus, forming a kind of neutral doublet. Such an atom would have very novel properties. Its external field would be practically zero, except very close to the nucleus, and in consequence it should be able to move freely through matter. Its presence would probably be difficult to detect by the spectroscope, and it may be impossible to contain it in a sealed vessel. On the other hand, it should enter readily the structure of atoms, and may either unite with the nucleus or be disintegrated by its intense field, resulting possibly in the escape of a charged H atom or an electron or both. ...

The existence of such nuclei may not be confined to mass 1 but may be possible for masses 2, 3, or 4, or more, depending on the possibility of combination between the doublets. The existence of such atoms seems almost necessary to explain the building up of the nuclei of heavy elements; for unless we suppose the production of charged particles of very high velocities it is difficult to see how any positively charged particle can reach the nucleus of a heavy atom against its intense repulsive field.

The discovery and isolation of the neutron, a discrete particle with a mass approximately that of a proton but without charge, was finally accomplished by another of Rutherford's colleagues, James Chadwick, in 1932. Several scientists had noticed that when beryllium was bombarded by α particles, the beryllium itself gave off radiation which was so powerful that it would knock hydrogen atoms out of a paraffin block. These workers thought that the radiation was electromagnetic in nature. Chadwick correctly identified the radiation as a stream of neutral particles, and through experimental efforts identified the mass of the particle as close to that of the proton. In a first brief statement, Chadwick explains his interpretations (11).

Possible Existence of a Neutron

It has been shown by Bothe and others that beryllium when bombarded by α-particles of polonium emits a radiation of great penetrating power . . .

Recently Mme. Curie-Joliot and M. Joliot found, when measuring the ionisation produced by this beryllium radiation in a vessel with a thin window, that the ionisation increased when matter containing hydrogen was placed in front of the window. The effect appeared to be due to the ejection of protons. . . .

I have made some experiments using the valve counter to examine the properties of this radiation excited in beryllium. . . .

These experiments have shown that the radiation ejects particles from hydrogen, helium, lithium, beryllium, carbon, air, and argon. The particles ejected from hydrogen behave, as regards range and ionising power, like protons with speeds up to about $3 \cdot 2 \times 10^9$ cm. per sec. The particles from the other elements have a large ionising power, and appear to be in each case recoil atoms of the elements. . . .

These results, and others I have obtained in the course of the work, are very difficult to explain on the assumption that the radiation from beryllium is a quantum radiation, if energy and momentum are to be conserved in the collisions. The difficulties disappear, however, if it be assumed that the radiation consists of particles of mass I and charge O, or neutrons. The capture of the α-particle by the Be^9 nucleus may be supposed to result in the formation of a C^{12} nucleus and the emission of the neutron. From the energy relations of this process the velocity of the neutron emitted in the forward direction may well be about 3×10^9 cm. per sec. The collisions of this neutron with the atoms through which it passes give rise to the recoil atoms, and the observed energies of the recoil atoms are in fair agreement with this view. . . .

It is to be expected that many of the effects of a neutron in passing through matter should resemble those of a quantum of high energy, and it is not easy to reach the final decision between the two hypotheses. Up to the present, all the evidence is in favour of the neutron, while the quantum hypothesis can only be upheld if the conservation of energy and momentum be relinquished at some point.

Later the same year Chadwick published the full account of his experiments (12).

It was shown by Bothe and Becker that some light elements when bombarded by α-particles of polonium emit radiations which appear to be of the γ-ray type. The element beryllium gave a particularly marked effect of this kind, and later observations by Bothe, by

Mme. Curie-Joliot and by Webster showed that the radiation excited
in beryllium possessed a penetrating power distinctly greater than
that of any γ-radiation yet found from the radioactive elements....

Quite recently, Mme. Curie-Joliot and M. Joliot made the very
striking observation that these radiations from beryllium and from
boron were able to eject protons with considerable velocities from
matter containing hydrogen. In their experiments the radiation from
beryllium was passed through a thin window into an ionisation vessel
containing air at room pressure. When paraffin wax, or other matter
containing hydrogen, was placed in front of the window, the ionisation in the vessel was increased, in some cases as much as doubled.
The effect appeared to be due to the ejection of protons,...

Chadwick calculated the amount of energy, in the form of electromagnetic quanta, necessary to produce these effects. The amount of
energy required was so large that it caused him to doubt the electromagnetic explanation.

Accordingly, I made further experiments to examine the properties of the radiation excited in beryllium. It was found that the
radiation ejects particles not only from hydrogen but from all other
light elements which were examined. The experimental results were
very difficult to explain on the hypothesis that the beryllium radiation was a quantum radiation, but followed immediately if it were
supposed that the radiation consisted of particles of mass nearly
equal to that of a proton and with no net charge, or neutrons....

This paper contains a fuller description of the experiments, which suggest the existence of neutrons and from which
some of the properties of these particles can be inferred....

Chadwick then presented the details of his experimental procedure.

The source of polonium was prepared from a solution of radium ... by deposition on a disc of silver. The disc had a diameter
of 1 cm. and was placed close to a disc of pure beryllium of 2 cm.
diameter, and both were enclosed in a small vessel which could be
evacuated, Figure 19. ...

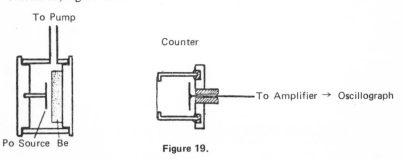

Figure 19.

When a sheet of paraffin wax about 2 mm. thick was interposed in the path of the radiation just in front of the counter, the number of deflections recorded by the oscillograph increased markedly. This increase was due to particles ejected from the paraffin wax so as to pass into the counter. By placing absorbing screens of aluminium between the wax and the counter the absorption curve ... was obtained. From this curve it appears that the particles have a maximum range of just over 40 cm. of air, assuming that an Al foil of 1·64 mg. per square centimetre is equivalent to 1 cm. of air. By comparing the sizes of the deflections (proportional to the number of ions produced in the chamber) due to these particles with those due to protons of about the same range it was obvious that the particles were protons. . . .

Chadwick determined the energy which the ejected protons possessed and then continued his experiments.

The effect of exposing other elements to the beryllium radiation was then investigated. . . . The element to be examined was fixed on a clean brass plate and placed very close to the counter opening. In this way lithium, beryllium, boron, carbon and nitrogen, as paracyanogen, were tested. In each case the number of deflections observed in the counter increased when the element was bombarded by the beryllium radiation. The ranges of the particles ejected from these elements were quite short, of the order of some millimetres in air. The deflections produced by them were of different sizes, but many of them were large compared with the deflection produced even by a slow proton. The particles therefore have a large ionising power and are probably in each case recoil atoms of the elements. Gases were investigated by filling the ionisation chamber with the required gas by circulation for several minutes. Hydrogen, helium, nitrogen, oxygen, and argon were examined in this way. Again, in each case deflections were observed which were attributed to the production of recoil atoms in the different gases. For a given position of the beryllium source relative to the counter, the number of recoil atoms was roughly the same for each gas. This point will be referred to later. It appears then that the beryllium radiation can impart energy to the atoms of matter through which it passes and that the chance of an energy transfer does not vary widely from one element to another. . . .

After calculating the quantum of electromagnetic radiation necessary to provide the observed recoil of various atoms, he found that the electromagnetic explanation was fraught with difficulties.

In general, the experimental results show that if the recoil atoms are to be explained by collision with a quantum, we must assume a larger and larger energy for the quantum as the mass of the struck atom increases.

The Neutron Hypothesis.—It is evident that we must either relinquish the application of the conservation of energy and momentum in these collisions or adopt another hypothesis about the nature of the radiation. If we suppose that the radiation is not a quantum radiation, but consists of particles of mass very nearly equal to that of the proton, all the difficulties connected with the collisions disappear, both with regard to their frequency and to the energy transfer to different masses. In order to explain the great penetrating power of the radiation we must further assume that the particle has no net charge. We may suppose it to consist of a proton and an electron in close combinations, the "neutron" discussed by Rutherford in his Bakerian Lecture of 1920.

Chadwick then provided calculations which were in close agreement with his experimental results.

When such neutrons pass through matter they suffer occasionally close collisions with the atomic nuclei and so give rise to the recoil atoms which are observed. Since the mass of the neutron is equal to that of the proton, the recoil atoms produced when the neutrons pass through matter containing hydrogen will have all velocities up to a maximum which is the same as the maximum velocity of the neutrons. The experiments showed that the maximum velocity of the protons ejected from paraffin wax was about $3 \cdot 3 \times 10^9$ cm. per second. This is therefore the maximum velocity of the neutrons emitted from beryllium bombarded by α-particles of polonium. From this we can now calculate the maximum energy which can be given by a colliding neutron to other atoms, and we find that the results are in fair agreement with the energies observed in the experiments. For example, a nitrogen atom will acquire in a head-on collision with the neutron of mass 1 and velocity $3 \cdot 3 \times 10^9$ cm. per second a velocity of $4 \cdot 4 \times 10^8$ cm. per second, corresponding to an energy of $1 \cdot 4 \times 10^6$ electron volts, . . .

Similarly, an argon atom may acquire an energy of $0 \cdot 54 \times 10^6$ electron volts, . . .

Both these values are in good accord with experiment.

It is possible to prove that the mass of the neutron is roughly equal to that of the proton, by combining the evidence from the hydrogen collisions with that from the nitrogen collisions. . . .

Chadwick provided an equation to describe the nuclear reaction which follows the collision of an α particle and a beryllium atom.

We have now to consider the production of the neutrons from beryllium by the bombardment of the α-particles. We must suppose that an α-particle is captured by a Be^9 nucleus with the formation of a carbon C^{12} nucleus and the emission of a neutron. The process is analogous to the well-known artificial disintegrations, but a neutron is emitted instead of a proton. The energy relations of this process cannot be exactly deduced, for the masses of the Be^9 nucleus and the neutron are not known accurately. It is, however, easy to show that such a process fits the experimental facts. We have

Be^9 + He^4 + kinetic energy of α
= C^{12} + n^1 + kinetic energy of C^{12} + kinetic energy of n^1. . . .

He then calculated the velocity of the neutrons from energy considerations and concluded:

> . . . it follows that the energy of emission of the neutron cannot be greater than about 8×10^6 electron volts. The velocity of the neutron must therefore be less than $3 \cdot 9 \times 10^9$ cm. per second. We have seen that the actual maximum velocity of the neutron is about $3 \cdot 3 \times 10^9$ cm. per second, so that the proposed disintegration process is compatible with observation. . . .

Chadwick then switched to the bombardment of boron with α particles. Since the mass of boron was better known his energy calculations would be more accurate.

> The mass of the beryllium nucleus has, however, not yet been measured, and, as was shown in §3, only general conclusions can be drawn from this reaction. Fortunately, there remains the case of boron. It was stated in §1 that boron bombarded by α-particles of polonium also emits a radiation which ejects protons from materials containing hydrogen. Further examination showed that this radiation behaves in all respects like that from beryllium, and it must therefore be assumed to consist of neutrons. It is probable that the neutrons are emitted from the isotope B^{11}, for we know that the isotope B^{10} disintegrates with the emission of a proton. The process of disintegration will then be
>
> $$B^{11} + He^4 \rightarrow N^{14} + n^1.$$
>
> The masses of B^{11} and N^{14} are known from Aston's measurements, and the further data required for the deduction of the mass of the neutron can be obtained by experiment. . . .

His energy calculations indicated the approximate mass of the neutron.

> Allowing for the errors in the mass measurements it appears that the mass of the neutron cannot be less than $1 \cdot 003$, and that it probably lies between $1 \cdot 005$ and $1 \cdot 008$. . . .

Chadwick then discussed some of the properties of the neutron.

> *The Passage of the Neutron through Matter.*—The electrical
> field of a neutron of this kind will clearly be extremely small . . .
>   In its passage through matter the neutron will not be
> deflected unless it suffers an intimate collision with a nucleus. . . .
>   Further, the neutron should be able to penetrate the nucleus
> easily, . . .

The ability to penetrate nuclei proved of great importance in future
years, as this makes the neutron as excellent atomic "bullet." There
are no electrostatic forces to repel it.

It is interesting to note that in his final remarks, Chadwick still
favors the hypothesis that the neutron is a combination of a proton
and an electron, and not a fundamental particle. New evidence later
caused him to reverse his position.

> It has so far been assumed that the neutron is a complex
> particle consisting of a proton and an electron. This is the simplest
> assumption and it is supported by the evidence that the mass of the
> neutron is about 1·006, just a little less than the sum of the masses
> of a proton and an electron. Such a neutron would appear to be the
> first step in the combination of the elementary particles towards the
> formation of a nucleus. It is obvious that this neutron may help us
> to visualise the building up of more complex structures, but the
> discussion of these matters will not be pursued further for such
> speculations, though not idle, are not at the moment very fruitful. It
> is, of course, possible to suppose that the neutron may be an ele-
> mentary particle. This view has little to recommend it at present,
> except the possibility of explaining the statistics of such nuclei as
> N^{14} . . . .
>   In conclusion, I may restate briefly the case for supposing that
> the radiation the effects of which have been examined in this paper
> consists of neutral particles rather than of radiation quanta. Firstly,
> there is no evidence from electron collisions of the presence of a
> radiation of such a quantum energy as is necessary to account for
> the nuclear collisions. Secondly, the quantum hypothesis can be
> sustained only by relinquishing the conservation of energy and
> momentum. On the other hand, the neutron hypothesis gives an
> immediate and simple explanation of the experimental facts; it is
> consistent in itself and it throws new light on the problem of nuclear
> structure.

> *Summary.*
>   The properties of the penetrating radiation emitted from
> beryllium (and boron) when bombarded by the α-particles of polo-

nium have been examined. It is concluded that the radiation consists, not of quanta as hitherto supposed, but of neutrons, particles
of mass 1, and charge 0.

Even though Chadwick didn't identify it as such, he had isolated the third of the major fundamental particles which produce the
gross structure of the atom. The number of protons and neutrons in
the nucleus produce the atomic weight of an atom and the number
of protons the atomic number. Although many fundamental particles
have since been discovered, the proton, neutron and electron still
form the basic structural units in our model of the atom.

Chapter 7

Final Comments

Between the years 1850 and 1932, man's ideas on the nature of the atom underwent drastic change. In 1850, the atom was considered a solid indivisible sphere, with all the atoms of an element identical. By 1932, all of these assumptions of Dalton had undergone revision. We saw that the discovery of the electron by J.J. Thomson altered the indivisibility postulate. The discovery and understanding of isotopes altered irrevocably the idea that all atoms of an element are identical. The discovery of the proton and the importance of atomic number altered the view that atomic weight was the critical property of the atom. Rutherford's nuclear model eliminated Dalton's solid sphere. A concise summary of the atomic model was presented by Aston in 1925 (2). To update this model to 1932, we must only add the neutron and its importance in determining nuclear structure and explaining atomic weight.

The atom is a kind of solar system with a positively charged nucleus for its sun and negative electrons rotating in orbits for its planets. But we must not forget that there are essential differences between the atomic system and our solar system. In the latter the orbits of the planets are all more or less in one plane so that the space occupied is in the form of a disk; in the normal atom, on the other hand, the movements of the electrons can take place in all planes so that as far as space occupied is concerned it does actually simulate a sphere, the form in which we have already represented it. More fundamental still is the fact that whereas the planets are held in their orbits by the gravitational pull of the sun, in the atom the movements of the electrons are controlled entirely by the electrical charge on the nucleus and not by its mass. Now the size of an atom will be the size of the orbits of its outermost planetary electrons; moreover, we find that all its chemical and spectroscopic properties depend on the movements of its system of planetary electrons. These depend on the electrical pull of the nucleus so that it is clear that the net positive charge on the nucleus is the most fundamental quality which determines the behaviour of an atom. To conform with our modern knowledge Dalton's postulate would now read: 'atoms of the same element are similar in properties *because they have the same nuclear charge.'*

In addition to causing modifications in Dalton's atomic model, the advances in our knowledge resulting from the discovery of X-rays and radioactivity presented completely new and unforeseen possibilities. By 1920, the possibility of harnessing the incredible energy involved in nuclear reaction was being considered. Einstein had demonstrated mathematically the equivalence of energy and matter in his famous equation $E = mc^2$ (where E is energy, m is mass and c is the speed of light). Since the speed of light is represented by very high numbers, a small amount of mass should convert into a large amount of energy. The weights of the atoms could be accurately determined by Aston's mass spectroscope, and it appeared that the weights of all nuclei larger than that of the hydrogen atom were smaller than the sum of their parts would indicate. Hence, it was suspected that some mass was converted into energy during the formation of atomic nuclei.

This book concludes with some of the speculations, first by Rutherford, and then by Aston, on the new and marvelous avenues of exploration the nuclear atom had opened, and some comments on the questions that were already being answered. Rutherford's statements are from a 1922 paper (46).

If, however, we take this structure of the helium nucleus as a working hypothesis, certain very important consequences follow. On the oxygen scale, the helium atom has a mass very nearly 4·000, while the hydrogen atom has a mass 1·0077. The mass of the helium atom is thus considerably less than that of four free H nuclei. Disregarding the small mass of the electrons, in the formation of 1 gram of helium from hydrogen there would be a loss of mass of 7·7 milligrams.

It is now generally accepted that if the formation of a complex system is accompanied by the radiation of energy E, the reduction of the mass m of the system is given by $E = mc^2$, where c is the velocity of light. This relation between mass and energy follows not only as a direct consequence of the theory of relativity, but can be derived directly from Maxwell's theory, as pointed out by Larmor. On this relation, the energy E liberated in the formation of 1 grm. of helium from hydrogen is equal to $6·9 \times 10^{18}$ ergs or $1·6 \times 10^{11}$ gramme-calories. This is an enormous amount of energy, large compared even with the total energy emitted during the complete disintegration of 1 grm. of radium and its products, namely, about $3\cdot7 \times 10^9$ gramme-calories. It can be calculated that the energy radiated in forming one atom of helium is equivalent to the energy carried by three or four swift α-particles from radium. On this view we can at once understand why it should be impossible to break up the helium nucleus by a collision with an α-particle. In fact, the helium atom should be by far the most stable of all the complex atoms.

It has been pointed out by Perrin and Eddington that in all probability the energy of radiation from our sun and the stars is derived mainly from the enormous emission of energy accompanying the formation of helium from hydrogen. If this be the case, it is easy to show that sufficient energy can be derived from this source for our sun to radiate at its present rate for several thousand million years, whereas the older theories of Kelvin and Helmholtz, in which the heat of the sun is ascribed to the gradual concentration of the material under gravity, make the life of the sun much shorter than modern estimates of the age of the earth and appear to be quite inadequate to provide the requisite energy.

This interesting suggestion of the probable origin of the greater part of the enormous energy radiated by the sun and stars is one of the first-fruits of the investigations on the structure of atoms. It is believed that the formation of helium from hydrogen occurs under certain conditions in the great central furnace of the sun and stars, but there is no evidence, so far, that this combination can be produced under laboratory conditions. It may be that it can be effected only under conditions of very high temperature and enormous intensity of radiation such as occur in the interior of a sun. Even then the process of formation may go on at a very slow rate and for periods measured by millions of years.

In the concluding excerpt, written in 1925 (2), Aston prophecies, among other things, the hydrogen bomb and ends this history on an optimistic, and hopefully justified, note.

The development of such artificial transmutation may turn out to be of enormous importance in connexion with atomic energy. The whole number rule is not mathematically exact, and it has been shown by direct measurement on the mass-spectrograph that an atom of helium which consists of four protons, two nuclear electrons, and two planetary electrons, weighs nearly 1 per cent. less than four atoms of hydrogen, each of which consists of one proton and one electron. The number of electrical particles is identical and the change of mass is ascribed to the different way they are arranged, and is called the packing effect. Hence if we could transmute hydrogen into helium mass would be destroyed and energy liberated. The quantity of energy is given at once by the theory of relativity and for any sensible quantity of matter transmuted is prodigious. Thus to transmute the hydrogen in a tumbler of water into helium would liberate energy sufficient to drive the Mauretania across the Atlantic and back at full speed. Transmutation of this kind is probably occurring in the interiors of the stars, but we are far indeed from effecting it in the laboratory. Nevertheless I have little doubt myself that man will one day be able to liberate and control this tremendous force, and I am optimist enough to believe that he will not devote it entirely to blowing his neighbours to pieces.

THE END

Brief Biographical Sketches
of the Major Characters

ASTON–Francis William Aston was born in Harborne, England, on September 1, 1877. He was educated at Cambridge University, from which he received a B.A. degree in 1912. After working for several years as a chemist for a brewery, and as a researcher at the University of Birmingham, he became a research assistant to J.J. Thomson at Trinity College, Cambridge University in 1910. He later worked with Lord Rutherford, and from 1921 to his death in 1945, was a researcher in the Cavendish Laboratory in Cambridge.

He is best known for his development of the mass-spectograph and for his discovery of nonradioactive isotopes. He received many awards and honors including the Nobel prize for chemistry in 1922.

BECQUEREL–Antoine Henri Becquerel was born in Paris, France on December 15, 1852. He earned his Doctor's degree in 1888. He was a lecturer at École Polytechnique from 1878-1892, and a Professor of Physics there from 1895 until his death in 1908.

He is best remembered for his extremely important discovery of radioactivity. Among his many awards was the 1903 Nobel prize in Physics which he shared with Marie and Pierre Curie.

BLACKETT–Patrick Maynard Stuart Blackett was born in London, England on November 18, 1897. He received an M.A. degree from Magdalene College, Cambridge University, in 1923. He was a Professor of Physics at Birkbeck College, London University, from 1933-1937, the University of Manchester from 1937-1953, and at Imperial College of Science and Technology from 1953-1965.

He is best known for his research with Rutherford on nuclear physics which included the isolation of the proton. He worked on radar and the atomic bomb during the second World War. Included among his many honors was the Nobel prize for Physics in 1948.

BOHR–Niels Henrik David Bohr was born in Copenhagen, Denmark on October 7, 1885. He received his Doctorate from the University of Copenhagen in 1911. He worked with J.J. Thomson at the Cavendish Laboratory in 1911-1912, and with Rutherford at Manchester in 1912-1913. He was a Lecturer in Physics at the University of Copenhagen in 1913-1914, and at

Victoria University in 1914-1916. He was Professor of Theoretical Physics at the University of Copenhagen from 1916 to his death in 1962.

He is best remembered for his work on the electron configurations of atoms, particularly in his model of the hydrogen atom. He was a founder of the Institute for Theoretical Physics in Copenhagen which he directed from 1920-1962. He received the Nobel prize for Physics in 1922.

CHADWICK—Sir James Chadwick was born in Manchester, Lancashire, England on October 20, 1891. He earned a Ph.D. degree from Cambridge University. He did extensive research at the Cavendish Laboratory with Rutherford and became Professor of Physics at the University of Liverpool in 1935.

He is well-known for his work with radioactivity, and particularly, for his discovery of the neutron. He worked on the development of the atomic bomb during the second World War. He was Knighted in 1945, and received the Nobel prize for Physics in 1935.

CROOKES—Sir William Crookes was born in London, England, on June 17, 1832. He attended the Royal College of Chemistry. He later received many honorary degrees. He founded and edited the *Chemical News* from 1859-1906.

His most important work was with cathode ray tubes. He determined that cathode rays were particulate in nature and developed a number of specialized cathode ray tubes which bear his name. He also did important work on radioactivity, and invented a method for detecting single alpha particle encounters with a zinc sulfide screen. He received numerous honors including Knighthood in 1897. He died in London in 1919.

CURIE—Marie Curie was born in Warsaw, Poland on November 7, 1867. She received her Doctorate in 1903 from Sorbonne. In 1906 she succeeded her husband as Professor of Physics, Sorbonne in Paris. She was the Director of the Radium Institute in Paris from 1914, until her death in 1934.

Her momentous achievements include the discovery of two radioactive elements, Radium and Polonium. Her work with radioactivity brought her two Nobel prizes, the first in 1903 with her husband Pierre and Becquerel, and the second, for Chemistry in 1911. Her death from leukemia was probably the result of the huge doses of radiation she was subjected to in her researches.

GEIGER—Johannes Wilhelm Geiger was born in Neustadt, Germany, on September 20, 1882. He received his Ph.D. degree from Erlangen University, Germany in 1906. He worked with Rutherford and Marsden on some of the most significant of the early nuclear physics research. He was an Instructor in Physics at the University of Manchester from 1906-1912. He became head of the Radioactivity Laboratory of the Reich Physics and Technology Institute, Berlin, in 1912 and became Professor of Physics at several German universities.

He is best known for his development, together with Muller, of the "Geiger" counter, for counting single alpha particles. His work which supported

Rutherford's nuclear model of the atom was very important. He died in Germany September, 1945.

MILLIKAN—Robert Andrews Millikan was born in Morrison, Illinois, on March 22, 1868. In 1895 he became the first person to receive a Ph.D. degree in Physics from Columbia University. He was a Professor of Physics at the University of Chicago from 1910 to 1921, and from 1921 through his retirement in 1945 was a professor at the California Institute of Technology.

His greatest accomplishment was his experimental determination of the fundamental unit of charge. He also investigated and named cosmic rays. Among his numerous awards was the Nobel prize in Physics in 1923. He died in California in 1953.

MOSELEY—Henry Gwyn Jeffreys Moseley was born in Weymouth, England, on November 23, 1887. He studied at Trinity College, Oxford, and studied with Rutherford in 1910-1911. He was a lecturer and then research fellow at Manchester. In 1913 he returned to Oxford.

His excellent work with X-ray spectra led him to formulate the concept of atomic number. This led to an improved understanding of the periodic table of the elements. His brilliant career was ended in 1915 with his death in the battle of Gallipoli, in Turkey.

ROENTGEN—Wilhelm Konrad Roentgen was born in Lennep, Prussia, on March 27, 1845. He received his Doctorate in Physics from Zurich, Switzerland, in 1869. He was a Professor at Wurzburg from 1885-1900, and at Ludwig-Maximillian University, Munich from 1900-1923. In 1888 he had become the Director of the Physical Institute at Wurzberg.

He is most famous for his discovery of X-rays. He received for this the first Nobel prize in Physics awarded in 1901. He died in Munich in 1923.

RUTHERFORD—Sir Ernest Rutherford was born in Spring Grove, near Nelson, New Zealand, on August 30, 1871. He studied with J.J. Thomson at Trinity College, Cambridge from 1895-1898. He was Professor of Physics at McGill University, Montreal, Canada from 1898-1907. He was at Victoria University, Manchester, England from 1907-1919. He left Manchester to succeed J.J. Thomson as Director of the Cavendish Laboratory, Cambridge, a post which he held until his death in 1937.

Rutherford is ranked by many with Newton as an experimental scientist. His accomplishments are too numerous to list here. Most important among these, however, are his theory of radioactive transmutation, his nuclear model of the atom, and his artificial transmutation of light elements by alpha particle bombardment. Every advancement in nuclear physics during the first quarter of the 20th century was influenced by him. He was a superb scientist. His incredible list of honors includes the Nobel prize for Chemistry in 1908. He was Knighted in 1914, and became Baron Rutherford of Nelson in 1931. He is buried in Westminster Abbey next to J.J. Thomson, near Newton.

SODDY—Frederick Soddy was born in Eastbourne, England, on September 2, 1877. He received a B.A. degree from Oxford in 1898 and an M.A. in 1910. He later received an LL.D. from Glasgow University. He was demonstrator in Chemistry at McGill University, Canada, from 1900-1902, during which time he collaborated with Rutherford in the study of radioactivity. He was at Aberdeen. University from 1914-1919, and Oxford University from 1919-1936.

He is best known for his work with Rutherford on the nature of radioactivity and for his work with isotopes. He received the Nobel prize in Chemistry in 1921. He died in Brighton, England in 1956.

THOMSON—Sir Joseph John Thomson was born in Chetham Hill, England, December 18, 1856. He received a B.A. degree from Trinity College, Cambridge in 1880, and became a Fellow of that college the same year. In 1884 he became Director of the Cavendish Laboratory, Cambridge, a post which he held until 1919.

He was a tremendously prolific researcher. His major accomplishments include the discovery and isolation of the electron, and the first coherent model of the atom involving subatomic particles (the "plum pudding" model). He developed the forerunner of the mass-spectroscope, and discovered isotopes in light elements. He received the Nobel prize in Physics in 1906, and was Knighted in 1908. He died in Cambridge, England in 1940. He is buried beside Rutherford in Westminster Abbey.

References

1. Aston, F.W., "Isotopes and Atomic Weights." Nature; July 15, 1920, vol. 105, pp. 617-619.
2. Aston, F.W., "The Structural Units of the Material Universe." Seventh Earl Grey Memorial Lecture: March 5, 1925, Oxford University Press, 1925. 23 pps.
3. Becquerel, H., *"Sur les radiations émises par phosphorescence."* Comptes rendus: 122, pp. 420-421, Feb. 24, 1896.
4. Becquerel, H., *"Sur les radiations invisibles émises par les corps phosphorescents."* Comptes rendus: 122, pp. 501-503, March 2, 1896.
5. Becquerel, H., *"Sur les radiations invisibles émises par les sels d'uranium."* Comptes rendus: 122, pp. 689-694, March 23, 1896.
6. Becquerel, H., *"Émission de radiations nouvelles par l'uranium métallique."* Comptes rendus: 122, pp. 1086-1088, May 18, 1896.
7. Blackett, P.M.S., "The Ejection of Protons from Nitrogen Nuclei, Photographed by the Wilson Method." The Proceedings of the Royal Society, A, vol. 107, 1925, pp. 349-360.
8. Bohr, N., "On the Constitution of Atoms and Molecules. Part I." Philosophical Magazine: ser. 6, vol. 26, July, 1913, pp. 1-25.
9. Bohr, N., "On the Constitution of Atoms and Molecules. Part II—Systems Containing only a Single Nucleus." Philosophical Magazine: ser. 6, vol. 26, Sept., 1913, pp. 476-502.
10. Bumstead, H.A., "The Heating Effects produced by Röntgen Rays in different Metals, and their Relation to the Questions of Change in the Atom." Philosophical Magazine: ser. 6, vol. 11, Feb., 1906, pp. 292-317.
11. Chadwick, J., "Possible Existence of a Neutron." Nature: Feb. 27, 1932, vol. 129, p. 312.
12. Chadwick, J., "The Existence of a Neutron." The Proceedings of the Royal Society, A, vol. 136, 1932, pp. 692-708.
13. Crookes, W., "On the Illumination of Lines of Molecular Pressure, and the Trajectory of Molecules." Philosophical Transactions of the Royal Society, Part I, 1879, pp. 135-164.
14. Curie, Mme., *"Rayons émis par les composés de l'uranium et du thorium."* Comptes rendus: 126, pp. 1101-1103, April 12, 1898.

15. Curie, P., and Curie, Mme., *"Sur une substance nouvelle radio-active, contenue dans la pechblende."* Comptes rendus: 127, pp. 175-178, July 18, 1898.

16. Curie, P., Curie, Mme., and Bémont, G., *"Sur une nouvelle substance fortment radio-active, contenue dans la pechblende."* Comptes rendus: 127, pp. 1215-1217, Dec. 26, 1898.

17. Demarcay, E., *"Sur le spectre d'une substance radio-active."* Comptes rendus: 127, p. 1218, Dec. 26, 1898.

18. Eve, A.S., "On the Secondary Radiation caused by β and γ Rays of Radium." Philosophical Magazine: ser. 6, vol. 8, Dec., 1904, pp. 669-685.

19. Geiger, H., and Marsden, E., "On a Diffuse Reflection of the α Particles." The Proceedings of the Royal Society, A, vol. 82, 1909, pp. 495-500.

20. Geiger, H., and Marsden, E., "The Laws of Deflection of α Particles through Large Angles." Philosophical Magazine: ser. 6, vol. 25, April, 1913, pp. 604-623.

21. Kelvin, Lord, "Aepinus Atomized." Philosophical Magazine: ser. 6, vol. 3, March, 1902, pp. 257-283.

22. Marsden, E., "The Passage of α Particles through Hydrogen." Philosophical Magazine: ser. 6, vol. 27, May, 1914, pp. 824-830.

23. Marsden, E., and Lantsberry, W.C., "The Passage of α Particles through Hydrogen -II." Philosophical Magazine: ser. 6, vol. 30, August, 1915, pp. 240-243.

24. Millikan, R.A., "A New Modification of the Cloud Method of Measuring the Elementary Electrical Charge, and the Most Probable Value of that Charge." Physical Review: vol. 29, Dec., 1909, p. 560.

25. Millikan, R.A., "A New Modification of the Cloud Method of Determining the Elementary Electrical Charge and the most Probable Value of that Charge." Philosophical Magazine: ser. 6, vol. 19, Feb., 1910, pp. 209-228.

26. Millikan, R.A., "The Isolation of an Ion, A Precision Measurement of its Charge, and the Correction of Stokes's Law." Physical Review: vol. 32, April, 1911, pp. 349-397.

27. Moseley, H.G.J., "The High-Frequency Spectra of the Elements." Philosophical Magazine: ser. 6, vol. 26, Dec., 1913, pp. 1024-1034.

28. Moseley, H.G.J., "The High-Frequency Spectra of the Elements. Part II." Philosophical Magazine: ser. 6, vol. 27, April, 1914, pp. 703-713.

29. Nagaoka, H., "Kinetics of a System of Particles illustrating the Line and the Band Spectrum and the Phenomena of Radioactivity." Philosophical Magazine: ser. 6, vol. 7, May, 1904, pp. 445-455.

30. Röntgen, W.C., "On a New Kind of Rays." Nature: Jan. 23, 1896, vol. 53, pp. 274-276.

31. Rutherford, E., "Uranium Radiation and the Electrical Conduction Produced by it." Philosophical Magazine: ser. 5, vol. 47, Jan., 1899, pp. 109-163.

32. Rutherford, E., "Excited Radioactivity and the Method of its Transmission." Philosophical Magazine; ser. 6, vol. 5, Jan., 1903, pp. 95-117.

33. Rutherford, E., "The Magnetic and Electric Deviation of the Easily Absorbed Rays from Radium." Philosophical Magazine: ser. 6, vol. 5, Feb., 1903, pp. 177-187.

34. Rutherford, E., "Some Remarks on Radioactivity." Philosophical Magazine ser. 6, vol. 5, April, 1903, pp. 481-485.

35. Rutherford, E., "Magnetic and Electric Deflection of the α Rays from Radium." Physical Review: vol. 22, Feb., 1906, pp. 122-123.

36. Rutherford, E., "Retardation of the α Particle from Radium in passing through Matter." Philosophical Magazine: ser. 6, vol. 12, August, 1906, pp. 134-146.

37. Rutherford, E., "The Mass and Velocity of the α Particles Expelled from Radium and Actinium." Philosophical Magazine: ser. 6, vol. 12, Oct., 1906, pp. 348-371.

38. Rutherford, E., "The Scattering of α and β Particles by Matter and the Structure of the Atom." Philosophical Magazine: ser. 6, vol. 21, May, 1911, pp. 669-688.

39. Rutherford, E., "The Structure of the Atom." Nature: Dec. 11, 1913, vol. 92, p. 423.

40. Rutherford, E., "The Structure of the Atom." Philosophical Magazine, ser. 6, vol. 27, March, 1914, pp. 488-498.

41. Rutherford, E., "Collision of α Particles with Light Atoms. I. Hydrogen." Philosophical Magazine: ser. 6, vol. 37, June, 1919, pp. 537- 561.

42. Rutherford, E., "Collision of α Particles with Light Atoms. II. Velocity of the Hydrogen Atoms." Philosophical Magazine: ser. 6, vol. 37, June, 1919, pp. 562-571.

43. Rutherford, E., "Collision of α Particles with Light Atoms. III. Nitrogen and Oxygen Atoms." Philosophical Magazine: ser. 6, vol. 37, June, 1919, pp. 571-580.

44. Rutherford, E., "Collision of α Particles with Light Atoms. IV. An Anomalous Effect in Nitrogen." Philosophical Magazine: ser. 6, vol. 37, June, 1919, pp. 581-587.

45. Rutherford, E., "Bakerian Lecture: Nuclear Constitution of Atoms." The Proceedings of the Royal Society, A, vol. 97, 1920, pp. 374-400.

46. Rutherford, E., "Electricity and Matter." Nature: August 5, 1922, vol. 110, pp. 182-185.

47. Rutherford, E., and Andrade, E.N.DA C., "The Wavelength of the Soft γ Rays from Radium B." Philosophical Magazine: ser. 6, vol. 27, May, 1914, pp. 854-868.

48. Rutherford, E., and Chadwick, J., "The Bombardment of Elements by α Particles." Nature: March 29, 1924, vol. 113, p. 457.

49. Rutherford, E., and Royds, T., "The Nature of the α Particle from Radioactive Substances." Philosophical Magazine: ser. 6, vol. 17, Feb., 1909, pp. 281-286.

50. Rutherford, E., and Soddy, F., "The Cause and Nature of Radioactivity, Part I." Philosophical Magazine: ser. 6, vol. 4, Sept., 1902, pp. 370-396.

51. Rutherford, E., and Soddy, F., "The Cause and Nature of Radioactivity, Part II." Philosophical Magazine: ser. 6, vol. 4, Nov., 1902, pp. 569-585.

52. Rutherford, E., and Soddy, F., "Radioactive change." Philosophical Magazine: ser. 6, vol. 5, May, 1903, pp. 579-591.

53. Soddy, F., "The Production of Radium from Uranium." Philosophical Magazine: ser. 6, vol. 9, June, 1905, pp. 768-779.

54. Soddy, F., "The Recent Controversy on Radium." Nature: Sept. 20, 1906, vol. 74, pp. 516-518.

55. Soddy, F., "Intra-atomic Charge." Nature: Dec. 4, 1913, vol. 92, pp. 399-400.

56. Thomson, J.J., "Cathode Rays." Philosophical Magazine: ser. 5, vol. 44, Oct., 1897, pp. 293-316.

57. Thomson, J.J., "On the Charge of Electricity carried by the Ions produced by Röntgen Rays." Philosophical Magazine: ser. 5, vol. 46, Dec., 1898, pp. 528-545.

58. Thomson, J.J., "On the Masses of the Ions in Gases at Low Pressures." Philosophical Magazine: ser. 5, vol. 48, Dec., 1899, pp. 547-567.

59. Thomson, J.J., "On the Structure of the Atom: an Investigation of the Stability and Periods of Oscillation of a number of Corpuscles arranged at equal intervals around the Circumference of a circle; with Application of the results to the Theory of Atomic Structure." Philosophical Magazine: ser. 6, vol. 7, March, 1904, pp. 237-265.

60. Thomson, J.J., "Some Further Applications of the Method of Positive Rays." Nature: May 29, 1913, vol. 91, pp. 333-337.

61. Thomson, J.J., "Positive Rays of Electricity." Nature: June 5, 1913, vol. 91, p. 362.

62. Townsend, J.S., "Electrical Properties of Newly Prepared Gases." Philosophical Magazine: ser. 5, vol. 45, Feb., 1898, pp. 125-151.

63. Van den Broek, A., "The Number of Possible Elements and Mendeléeff's 'Cubic' Periodic System." Nature: July 20, 1911, vol. 87, pp. 78.

64. Van den Broek, A., "Intra-atomic Charge." Nature: Nov. 27, 1913, vol. 92, pp. 372-373.

65. Van den Broek, A., "Intra-atomic Charge and the Structure of the Atom." Nature: Dec. 25, 1913, vol. 92, pp. 476-478.

66. Wilson, H.A., "A Determination of the Charge on the Ions produced in Air by Röntgen Rays." Philosophical Magazine: ser. 6, vol. 5, April, 1903, pp. 429-441.

Index